DYNAMIC
DEVELOPMENT

SHIFTING
DEMOGRAPHICS

CHANGING
DIETS

The story of the rapidly evolving food system in Asia and the Pacific
and why it is constantly on the move

FOOD AND AGRICULTURE ORGANIZATION OF THE UNITED NATIONS
Bangkok, 2018

Required citation:
FAO. 2018. *Dynamic development, shifting demographics, changing diets.* Bangkok. 172 p. Licence: CC BY-NC-SA 3.0 IGO.

COVER PHOTOGRAPH ©Shutterstock/CherylRamalho

CONTENTS

Boxes, figures and maps

Boxes

Figures

Maps

FOREWORD

Food – it is the foundation that humanity relies upon for survival – and yet its availability is often taken for granted. The general assumption, at least in the most affluent parts of the world, is that food will always be around, precisely because we need it for our survival. However, while the world presently produces enough food for everyone (although too many do not receive their fair share), food security in the coming decades is far less certain. Because of this uncertainty, and because the world still has more than 800 million undernourished people, eradicating food insecurity and malnutrition has been included as one of the Sustainable Development Goals (SDGs), i.e. SDG2 – the Zero Hunger goal.

While SDG2 is a universal goal, agreed to by all United Nations (UN) Member States, the context in which countries strive to achieve it differs from place to place. One size does not fit all. Nowhere is this more evident than in Asia and the Pacific, a region that is different from the rest of the world in many ways. The pace of economic growth and urbanization has been extremely rapid in the region, and is expected to remain that way for the near future. This context of rapid change has a range of implications for food systems and provides the motivation for writing a publication that focuses on the Asia-Pacific region.

Economic growth has led to increased income for the poor and large reductions in poverty rates, with the region having achieved the first Millennium Development Goal (MDG) target of reducing poverty incidence by half between 1990 and 2015. Unfortunately, most of the growth in incomes has accrued to those in the top echelon of the income distribution, widening the gap between the affluent and the poor, which will make it more difficult to achieve many of the SDGs. Economic growth has also led to a decline in fertility rates, leading to ageing of the population. This change is playing out at a much faster pace than historically has been the case in most developed countries, implying that the region has far less time to build the financial infrastructure and social security systems needed to address the consequences of an ageing population effectively.

By 2021, more than half of the region's population will live in urban areas, due to both rural to urban migration and the expansion of urban areas. This urbanization is leading to a substantial change in the organization of food systems in terms of production, marketing and consumption. Supply chains are becoming longer geographically, but often with fewer intermediaries, and are becoming more demanding in terms of quality, food safety and traceability. These demands of the private sector and by consumers can be difficult for farmers, especially smallholders, to meet. At the same time, urban food demand is a burgeoning opportunity for rural producers to increase their incomes and become more prosperous.

Urbanization is also modifying the settings where food is made available and accessible for purchase and consumption. These 'food environments' determine what types of food consumers can access at a given moment in time, at what price and with what degree of convenience and desirability, thus greatly influencing their dietary intake and nutritional outcomes.

Economic growth is enabling consumers to diversify their diets, and as a result, non-staple foods now account for well over half of total food expenditures. This means that farmers must change what they produce, shifting towards more protein-rich foods as well as more fruits and vegetables. However, the availability and consumption of heavily processed foods that are high in salt, sugar or fat are also increasing substantially. Coupled with reduced physical exercise, these changing consumption patterns are causing overweight and obesity rates to climb in all countries – they have already reached extremely high levels in Pacific Island countries. In turn, the toll from non-communicable diseases (NCDs) is rising and straining public health systems.

The increased demands for dietary diversification and food in general (due to continued population growth) have also contributed, along with other factors, to increasing stress on our natural environment. These stresses include adverse impacts on land and water resources, climate change, an increased frequency of natural disasters and damage to human and ecosystem health. Not only do these stresses affect our overall quality of life, they also make it more difficult to produce food in a sustainable manner for future generations.

In addition to diversifying their diet, consumers are also spending much of their extra income on non-food items such as health, education, clothing, transport, entertainment and consumer electronics. These shifts in spending patterns are driving a structural transformation of regional economies and rural livelihoods – approximately 90 percent of rural households in a group of large Asian countries earn money from non-agricultural activities.

New technologies are spreading rapidly – many people are now referring to the emergence of a fourth Industrial Revolution that combines physical and biological systems with the digital world. Emerging technologies (4G, broadband, the Internet of Things, smart phones, remote sensing, artificial intelligence, drones and sensor networks), as well as more traditional technologies such as mechanization, will affect how we grow our food and manage our natural environment. New technologies such as blockchain will affect how food moves from one location to another, with great promise for bringing greater accountability and transparency in food traceability, as well as in supporting smallholders in accessing emerging markets. Looking further ahead, the fourth Industrial Revolution may also affect where we grow our food, as urban and laboratory agriculture are becoming increasingly popular.

Understanding all of these trends, their interdependencies and their implications was not an easy task, nor was it linear. The process involved consulting a range of internal and external experts, numerous conversations over lunch and coffee, commissioning papers on specialized topics, presenting interim analysis and results at different seminars and experimenting with different written formats. The end result is this publication that examines what these trends facing the Asia-Pacific region will mean for food systems and a wide variety of stakeholders – governments, the private sector, consumers and smallholder farmers, fishers, herders and foresters. Its main purpose is to synthesize and integrate trends and developments across a wide range of topics, many of which are outside the traditional boundaries of food policy discussions. I hope that it will contribute to more informed food policy debates in Asia-Pacific countries in the context of rapid change and more complex food systems.

Kundhavi Kadiresan
Assistant Director-General and Regional
Representative for Asia and the Pacific

Dr Tim Martyn worked tirelessly to improve food security and nutrition in the Pacific. He was a committed young professional with talent, insight and vision, and inspired many of us. Everyone is profoundly saddened by his death and he will be greatly missed.

Acknowledgements

This publication benefited from inputs, comments and suggestions from a wide range of people in FAO's Regional Office for Asia and the Pacific (RAP), FAO's subregional office for the Pacific (SAP), FAO country offices in the region and FAO headquarters. A number of external reviewers also provided very helpful feedback at various stages.

The report was written under the leadership and technical supervision of David Dawe with strong support from a core technical team that consisted of Vinod Ahuja, Sunniva Bloem, Beau Damen, Tim Martyn and Louise Whiting (all from FAO's Regional or Subregional Offices in the Asia-Pacific region). Strategic guidance and supervision was provided by Xiangjun Yao, Regional Programme Leader in RAP and overall supervision by Kundhavi Kadiresan, FAO Assistant Director General and Regional Representative for Asia and the Pacific. Surawishaya Paralokanon provided a range of essential inputs too numerous to mention including research, preparation of graphs, management of references, coordination of the publishing process and various administration services.

Key external reviewers included Steven Jaffee, Lead Economist at the World Bank; Peter Timmer, Cabot Professor of Development Studies, emeritus, Harvard University; and Keith Wiebe, Senior Research Fellow at the International Food Policy Research Institute (IFPRI). They helped to shape the publication and the core team is extremely thankful for their contributions.

Sridhar Dharmapuri, Mayling Flores Rojas, Eva Gálvez Nogales and Gerard Sylvester from FAORAP wrote boxes or case studies for the concluding chapter. Colleagues at RAP who provided useful comments, inputs and assistance at different stages were Lois Archimbaud, Anthony Bennett, Srijita Dasgupta, Allan Dow, Thomas Hofer, Melina Lamkowsky, Rachele Oriente, Clara Park, Maria Paula Sarigumba, Caroline Turner and participants at various internal seminars. SAP colleagues who provided helpful inputs were Louison Dumaine Laulusa, Eriko Hibi, Joseph Nyemah, Shukrullah Sherzad and Anna Tiraa.

Colleagues at FAO country offices in the region who provided useful inputs include Lalita Bhattacharjee (Bangladesh), Nina Brandstrup (Sri Lanka), Jose-Luis Fernandez, Maria Pastores, Maria Quilla (the Philippines) and Mark Smulders (Indonesia), as well as participants from all country offices in the annual regional management meetings in Bangkok.

From FAO headquarters, colleagues from multiple technical divisions and strategic programmes provided valuable comments and suggestions: Angela Bernard, Karel Callens, Benjamin Davis, Juan Garcia Cebolla, Fatima Hachem, Adriana Ignaciuk, Jessica Fanzo, Daniela Kalikoski, Erdgin Mane, Jamie Morrison, You Ny, Ahmed Raza, Jodean Remengesau, Marco Sánchez Cantillo, Cassandra Walker, Emilie Wieben, Ramani Wijesinha Bettoni and Trudy Wijnhoven. We apologize to people whose names we missed due to quirks of the document sharing system that recorded their suggestions as coming from "Guest."

Last but not least, this publication would not have been possible without a helpful financial contribution from the International Fund for Agricultural Development (IFAD). FAORAP gratefully acknowledges that contribution. In addition, several of their staff provided insightful comments and suggestions at various stages, in particular Fabrizio Bresciani, Thomas Chalmers and Dilva Terzano.

Editing was provided by Robin Leslie, and layout was done by QUO Global in Bangkok. For more information about this publication, please contact FAO-RAP@fao.org.

Acronyms and abbreviations

ADB	Asian Development Bank
AFOLU	Agriculture, Forestry and Land Use
AMR	antimicrobial resistance
ASEAN	Association of Southeast Asia Nations
CPI	consumer price index
DRRM	disaster risk reduction and management
FAO	Food and Agriculture Organisation
GDI	Gender Development Index
GDP	gross domestic product
GHG	greenhouse gas
GIs	geographical indications
IFAD	International Fund for Agricultural Development
IFOAM	International Federation of Organic Agriculture Movements
IFPRI	International Food Policy Research Institute
NACCFL	Nepal Agricultural Cooperative Central Federation
NCD-RisC	NCD Risk Factor Collaboration
NCDs	non-communicable diseases
NGO	non-governmental organization
OECD	Organisation for Economic Co-operation and Development
PGS	Participatory Guarantee Systems
PICs	Pacific Island Countries
PPP	purchasing power parity
SDGs	Sustainable Development Goals
SINER-GI	Strengthening INternational Research on Geographical Indications
SLR	sea level rise
SPIS	Solar powered irrigation systems
UN	United Nations
UNDESA	United Nations Department of Economic and Social Affairs
UNDP	United Nations Development Programme
UNESCAP	United Nations Economic and Social Commission for Asia and the Pacific
UNICEF	United Nations Children's Fund
USDA	United States Department of Agriculture
WEAI	Women's Empowerment in Agriculture Index
WFP	World Food Programme
WHO	World Health Organization
WWAP	United Nations World Water Assessment Programme

Country groupings used in this report

East Asia
China
Mongolia
Democratic People's Republic of Korea

Southeast Asia
Cambodia
Indonesia
Lao People's Democratic Republic
Malaysia
Myanmar
Philippines
Thailand
Timor-Leste
Viet Nam

South Asia
Afghanistan
Bangladesh
Bhutan
India
Maldives
Nepal
Pakistan
Sri Lanka

Pacific
American Samoa
Cook Islands
Fiji
French Polynesia
Guam
Kiribati
Marshall Islands
Micronesia (Federated States of)
Nauru
New Caledonia
Niue
Northern Marianas
Papua New Guinea
Samoa
Solomon Islands
Tokelau
Tonga
Tuvalu
Vanuatu
Wallis and Futuna Islands

High-income countries
Australia
Brunei Darussalam
Japan
New Zealand
Palau
Republic of Korea
Singapore

1

INTRODUCTION

The Asia-Pacific region is undergoing rapid change and the food systems[1] that produce and deliver food to its consumers are not exempt from these trends. Some of the challenges confronting these systems are longstanding and generally familiar, but remain important nonetheless. For example, regional and global populations continue to grow and concomitantly the demand for food, placing further stresses on our already heavily burdened natural resources.

But in addition to these challenges that have existed since at least the time of Malthus, rapid change has brought with it newer trends in regional food systems. Sustained economic growth has led to ageing populations, greater urbanization, increased international trade and a structural transformation of regional economies, all of which have profound implications for food systems. Ageing populations, especially in rural areas, have implications for labour supply and technology adoption. Greater urbanization has led to reduced physical activity and an increased demand for convenience that has increasingly important implications for nutritional outcomes such as obesity. This urbanization has also led to an increased reliance of farm households on remittances and in some cases, a feminization of farming. Increased international trade due to lower costs of transportation and communication and lower trade barriers has increased the variety of foods available to consumers and placed additional pressures on farmers to be competitive. Structural transformation towards other sectors of the economy (industry and services) has deeply affected the livelihood strategies of farm households and the time they spend on farming. All of these changes have taken place against a background of increased inequality, rapid growth in the spread of information and communication technologies and a changing climate.

---

1   The term 'food systems' covers the entire range of actors and their interlinked value-adding activities involved in the production, aggregation, processing, distribution, preparation, consumption and disposal of food products that originate from agriculture, forestry or fisheries, and parts of the broader economic, societal and natural environments in which they are embedded (FAO, 2018c).

Collectively, the above trends mean that food systems (and thus, food policy) are becoming substantially more complex compared to the past. Interactions between different components of the food systems are increasingly becoming more intricate and multidimensional, linking sectors and actors in unprecedented and sometimes novel ways. Agriculture is now just one part of an integrated, more science- and capital-intensive, globalized food system. The future will see further profound changes in what is grown, how it is grown, where it is grown and how it moves from one place to another. Given this reality, this publication considers interactions among different sectors and different spaces, the essence of the food systems approach.

Thus, food systems now fall within the purview of different line ministries and different stakeholders outside of government (e.g. academia, civil society organizations, the private sector). In order for food policies to be effective, it is important that all stakeholders are aware of and understand the trends not only within agriculture, but also across environmental, health and nutrition, urban planning, finance and trade domains.

The trends noted above do not take place in a sociocultural vacuum, so the implications of these trends depend upon the specific context in which they occur. This publication focuses on the Asia-Pacific region. While far from homogeneous, this region does have certain characteristics that distinguish it from other major world regions. First, it has a large population, accounting for more than half (55 percent) of the world's total. The region now accounts for 41 percent of global gross domestic product (GDP) measured in purchasing power parity or 'PPP' terms, up from 25 percent in 1990. Further, most of the region's population is concentrated in one or another of its larger countries – 92 percent of the population lives in a country with a population of more than 50 million people (these percentages are much lower in Africa, Latin America and Europe). Second, the growth of GDP per capita has been, by far, more rapid than elsewhere in the world. Since 1990, and despite the Asian financial crisis in 1997/1998, GDP per capita grew at an annual average rate of 6.5 percent, while no other region had a rate above 3 percent (Figure 1.1). Third, the pace of urbanization in Asia and the Pacific has been the most rapid in the world, increasing from 30 percent in 1990 to 47 percent in 2016 (Figure 1.2). The increase in urbanization is projected to remain the fastest of any continent in the world between now and 2050.

Figure 1.1 Annual average growth of GDP per capita by decade by region

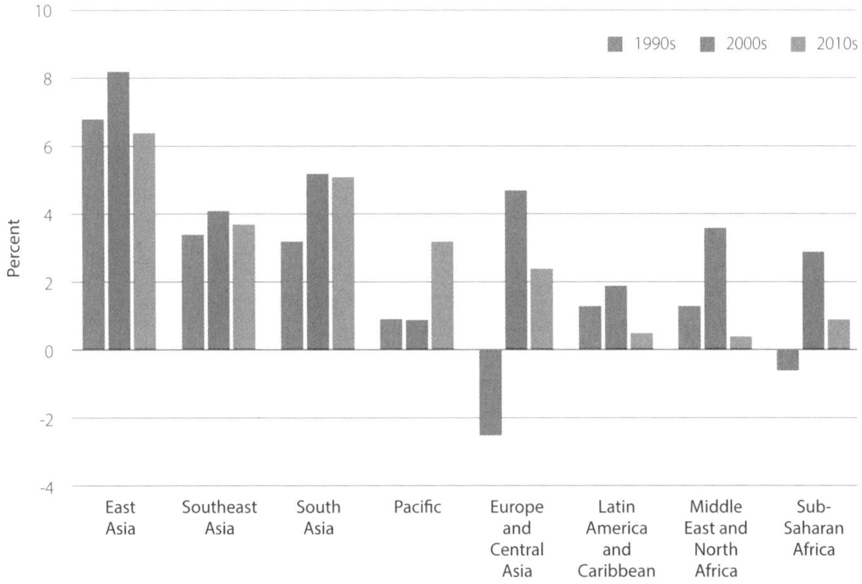

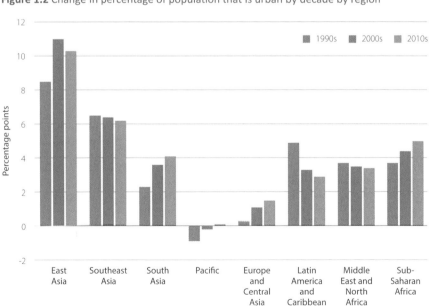

Source of raw data: World Bank (2018)
Note: High-income countries excluded from all regions.

Figure 1.2 Change in percentage of population that is urban by decade by region

Source of raw data: World Bank (2018)
Note: High-income countries excluded from all regions. Estimates for 2010s based on data until 2016, scaled up from six years to ten years.

Figure 1.3 Agricultural area per capita by continent or subregion

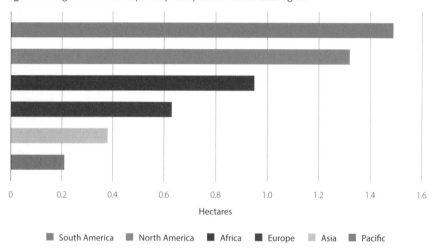

Source of raw data: FAO (2018b)
Note: Australia (15.4 hectares per capita) not shown to make differences among the other continents more visible.

Natural resource endowments in Asia and the Pacific also differ from elsewhere in the world, particularly regarding land. Population densities are very high, and as a result, agricultural area per capita is lower in Asia than on any other continent, with that on the Pacific Islands being even lower (Figure 1.3). The low agricultural area per capita translates into small farm sizes and a preponderance of smallholder farmers. Similar to sub-Saharan Africa, about 95 percent of all farms in Asia and the Pacific are less than 5 hectares in size, compared to about 50 percent in Latin America and the Caribbean (Lowder, Skoet and Singh, 2014).

In terms of water, a recent report from the Organisation for Economic Co-operation and Development (OECD) names northwest India and north China as two of the world's top three hotspots in terms of water-related risks to food production (OECD, 2017). At the same time, more than in other parts of the world, many countries tend to be reliant on irrigation systems, with a much larger percentage of cultivated area in Asia equipped for irrigation (41 percent) than in any other continent (Portmann, Siebert, & Döll, 2010) (Map 1.1). Indeed, North America ranks second at just 13 percent, far behind. Thus, the challenges for water management are substantially different in the region than in other parts of the world.

The Asia-Pacific region is also different from the rest of the world in another way – its linguistic and socio-economic diversity.[2] For example, economic growth has been rapid throughout Asia (and is projected to continue), while it has been much slower in the Pacific countries (Figure 1.1). In terms of urbanization (Figure 1.2), the increases have been the most rapid in East and Southeast Asia, with the pace being slower in South Asia and no trend being evident in the Pacific. Land is scarce throughout the region, but water scarcity is more variable. For example, in Southeast Asia and the Pacific, all countries (except for Singapore) have renewable water resources per capita that are above the median for the world (FAO, 2018a), so irrigation systems are generally less widespread than in East and South Asia (Map 1.1).

Map 1.1 Percentage of cultivated area equipped for irrigation

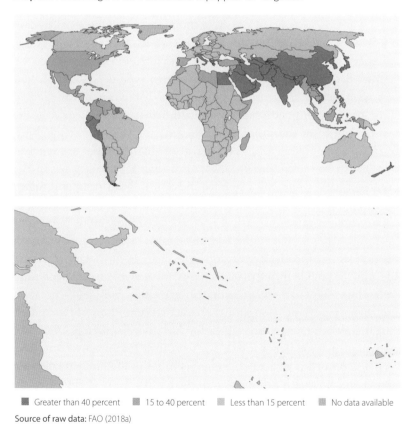

■ Greater than 40 percent   ■ 15 to 40 percent   ■ Less than 15 percent   ■ No data available

Source of raw data: FAO (2018a)

---

2 In most of this publication, the Asia-Pacific region is divided into five subregions: East Asia, the Pacific, South Asia, Southeast Asia and high-income economies within those four geographic subregions (although data for the last of those subregions will not always be reported). Occasionally, high-income countries are included in their respective geographical subregions, e.g. Japan and Republic of Korea in East Asia. When this is done, it is explicitly noted. For a list of the countries contained in each of these groupings, please see page xiii. Central and Western Asia are not included as part of Asia and the Pacific in this report.

Chapter 2 of this report examines the economic, sociocultural and biophysical trends discussed briefly above in more detail. It provides analysis on the mega trends of economic growth alongside growing income inequality, other types of inequality (gender, rural vs. urban), population growth and demographic shifts, urbanization and migration, globalization and international trade and the structural transformations that occur in societies as a consequence.

Chapter 3 assesses how these trends, particularly economic growth and population growth, are affecting the region's natural resource base in terms of land and water, while also discussing the impact of climate change on food production. These trends, in conjunction with those discussed in Chapter 2, characterize the broad environment within which regional food systems are changing. An understanding of these trends is essential for informed food policy.

These socio-economic and environmental mega trends have important implications for food systems. While it is common to use the shorthand of 'farm to fork' to discuss food systems, this book inverts that order and begins with consumers and consumption. Market economies are demand-driven, so it is consumer demand that drives the entire system – farmers will soon find themselves out of business if they do not produce what consumers are demanding.

Chapter 4 begins with a discussion on malnutrition in the region in the context of the nutrition transition, and broadens the focus to include both undernutrition and overnutrition. Perhaps surprisingly in light of the rapid economic growth and accumulation of wealth that has taken place in the region, undernutrition remains a serious problem in both rural and urban areas, with the prevalence of stunting still at high levels in many countries. At the same time, overweight and obesity are becoming more important over time, presenting a challenge to public health systems. Following on from this discussion, the chapter explores the changes in dietary structure that underpin the trends in nutritional status, with staple foods becoming relatively less important to diets and animal source foods, fruits and vegetables becoming more important, alongside increased consumption of foods high in fats, sugar and salt. Dietary diversification has thus liberated people from an excessive reliance on staple foods, with positive implications for some nutritional outcomes but with new challenges in terms of managing obesity. The chapter concludes with a discussion of the trends that underpin shifting diets – growing incomes, food prices, urbanization and diversification within the agriculture sector. Shifting diets challenge farmers to meet these emerging demands.

Chapter 5 describes the transformation of value chains in the region that are delivering the food produced by farming systems (described in Chapter 6). These value chains are becoming increasingly sophisticated and complex, supported by an expansion of infrastructure – roads, electricity, mobile telephones and access to the Internet. Strong demand for more convenient food preparation is shaping these value chains, leading to an increasing diversity of retail market outlets. Food manufacturing is growing in importance, as is international trade and globalization.

Chapter 6 looks at trends in rural livelihoods and demographics, documenting the declines in rural poverty and the growing importance of non-farm income, as well as the ageing of rural populations and the feminization of agricultural employment in some parts of the region. It also examines the causes and implications of trends in farming systems, including greater labour productivity, rising rural wages and mechanization, and changing farm sizes. It emphasizes in particular the critical importance of higher agricultural labour productivity for sustainable rural poverty alleviation.

Given the increasing complexity of food systems and the heterogeneous contexts in which these systems function, Chapter 7 does not attempt to provide solutions to the wide range of issues facing the evolving food systems of the region. Rather, it describes a few specific examples of how different stakeholders are responding to these trends in various ways, including the advantages and disadvantages of such approaches. The advantages and disadvantages will differ not only among countries, but also within countries. These examples span the natural environment, nutrition, value chains and farmers' livelihoods, but do not pretend to provide a comprehensive snapshot of developments in the region. We are aware that many other initiatives are also taking place and merely hope that the examples presented here give some small flavour of the myriad developments that are taking place in response to the trends described in this report.

References

FAO. 2018a. *AQUASTAT* [online]. http://www.fao.org/nr/water/aquastat/main/index.stm

FAO. 2018b. *FAOSTAT* [online]. www.fao.org/faostat/

FAO. 2018c. Food systems and value chains: definitions and characteristics. In: Production and Resources: Developing Sustainable Food Systems and Value Chains for Climate-Smart Agriculture [online]. http://www.fao.org/climate-smart-agriculture-sourcebook/production-resources/module-b10-value-chains/chapter-b10-2/en/

Lowder, S.K., Skoet, J. & Singh, S. 2014. What do we really know about the number and distribution of farms and family farms in the world? ESA *Working Paper*, 14(2). (also available at http://www.fao.org/family-farming/detail/en/c/281544/).

Organisation for Economic Co-operation and Development (OECD). 2017. *Water risk hotspots for agriculture.* (also available at http://www.oecd-ilibrary.org/agriculture-and-food/water-risk-hotspots-for-agriculture_9789264279551-en).

Portmann, F.T., Siebert, S. & Döll, P. 2010. MIRCA2000—Global monthly irrigated and rainfed crop areas around the year 2000: A new high-resolution data set for agricultural and hydrological modeling. *Global Biogeochemical Cycles*, 24(GB1011). https://doi.org/10.1029/2008GB003435

World Bank. 2018. *World development indicators* [online]. https://data.worldbank.org/products/wdi

CHAPTER 2

MEGA TRENDS
IN ASIA AND THE PACIFIC

Economic growth

The countries of the Asia-Pacific region are diverse in terms of income levels (Map 2.1), but have generally experienced more rapid economic growth over the past few decades compared to the rest of the world. From 2000 to 2016 the region had an average annual GDP per capita growth rate of 5 percent, compared to the world average of 2 percent. Among its four subregions, East Asia has experienced the most rapid growth in GDP per capita since the start of the century in percentage terms, followed not far behind by Southeast Asia and South Asia (which has been a relative newcomer to sustained rapid growth; see Figure 2.1). This growth has been driven by many factors, including stable macroeconomic policies that encourage investment, relative sociopolitical stability, a more educated labour force with more participation by women, a demographic dividend (where there is a bulge in the share of the population that is of working age) and rapid urbanization.

The Pacific has lagged behind in terms of economic growth, which has had important implications for agriculture, poverty and nutrition (see Box 2 in Chapter 4). Nearly all countries in the Pacific have experienced very slow growth, with the overall growth rate for the region remaining at less than 2 percent (see Figure 2.1).

Map 2.1 GDP (PPP) per capita in Asia and the Pacific

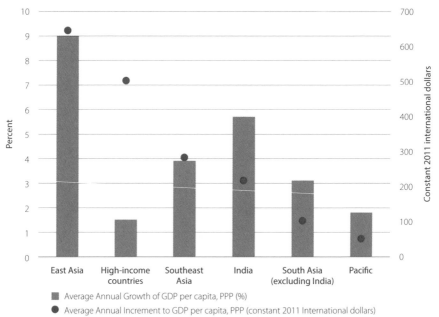

Source of raw data: World Bank (2016)

Figure 2.1 Absolute and percentage growth in GDP (PPP) per capita by subregion, 2000–2015

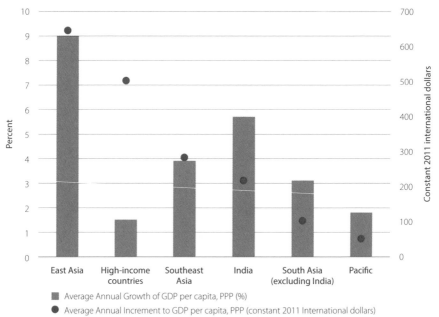

Source of raw data: World Bank (2018)

Figure 2.2 Annual average percentage growth in incomes of the poorest and richest quintiles, early 2000s to early 2010s

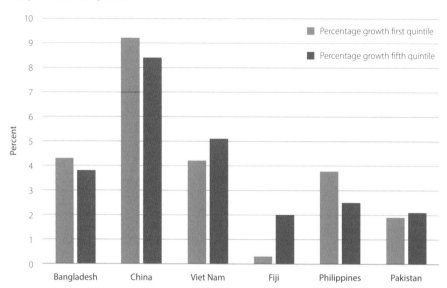

Source of raw data: World Bank (2018)

Note: Initial and final years for each country are as follows: Bangladesh (2000–2010), China (2008–2012), Viet Nam (2002–2014), Fiji (2002–2008), Philippines (2000–2012), Pakistan (2001–2013).

Economic growth can play an important role in poverty alleviation and improved food security and nutrition, provided it reaches the poor. In most countries of the region, the income of the poorest 20 percent has grown rapidly in percentage terms, at rates roughly comparable to growth in the income of the wealthiest 20 percent (see selected countries in Figure 2.2). A recent World Bank analysis (World Bank, 2016a) found that, among 11 Asian countries analysed, the income of the bottom 40 percent increased in all cases. Thus, the poor are benefiting from economic growth. As a result of this broad distribution of growth, there has been a widespread decline in extreme poverty rates. Indeed, most of the decline in global extreme poverty since 1990 has been due to declines in the Asia-Pacific region (World Bank, 2016a).

Inequality

In nine of the eleven countries analysed in the region, the percentage increase in income for the bottom 40 percent in the past few years (2008–2013) was greater than the average percentage increase in income for the entire population (World Bank, 2016a). Changes in Gini coefficients, a common measure of inequality, do not show substantial change over time in East Asia and the Pacific, although they have increased over time in South Asia (World Bank, 2016a). According to these measures, inequality does not seem to be getting appreciably worse.

But a constant Gini coefficient can be misleading in terms of understanding how most of the additional income is distributed.[1] In fact, most of the increased income within countries of the region has accrued to the upper reaches of the income distribution (the red bars are bigger than the blue bars in the selected countries of Figure 2.3). This phenomenon is similar to what is happening across countries. While the high-income countries in the region have grown slowly in percentage terms, their growth in absolute terms (i.e. in US dollars or PPP dollars per capita) has been larger than anywhere else except East Asia, as their small percentage growth is applied to a very large starting income (see the circles in Figure 2.1). Thus, absolute income gaps between developed and developing countries in the region are actually growing despite the fact that percentage growth rates are generally slower in high-income countries.

Overall, income is distributed very unequally, with the poorest 10 percent of the population receiving less than 3 percent of the total income in many countries (Map 2.2). In India, the top 10 percent now earns about 55 percent of the national income, nearly as extreme as in the Middle East, which is the most unequal of all major world regions (Alvaredo et al., 2018).

---

1   Gini coefficients will remain unchanged if all people have equal percentage rates of income growth. Imagine two people with incomes of USD 1 000 and USD 100 respectively, and there is an increase in total income of USD 220, with USD 200 going to the rich party and USD 20 to the poor party. If the Gini coefficient is used as the measure of inequality, then inequality will be measured as remaining constant even though the richer person received 91 percent of the additional income. This is because the rich person had 91 percent of the initial income. Thus, a constant Gini coefficient does not mean that additional income is being shared equally.

Figure 2.3 Annual average absolute growth in incomes of the poorest and richest quintiles, early 2000s to early 2010s

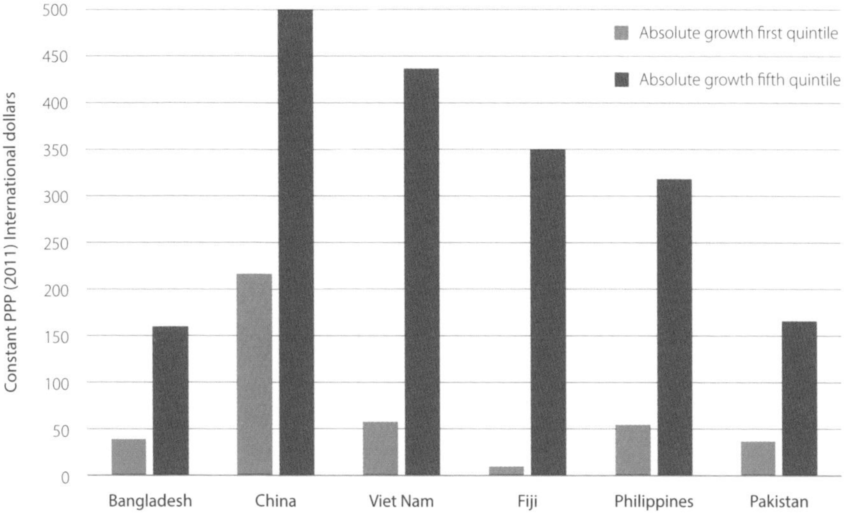

Source of raw data: World Bank (2018)

Note: Initial and final years for each country are the same as in Figure 2.2; vertical axis is truncated at 500 international dollars in order to show all countries more clearly. Actual value for the top quintile in China is I$1805.

Map 2.2 Share of GDP per capita accruing to the poorest 10 percent

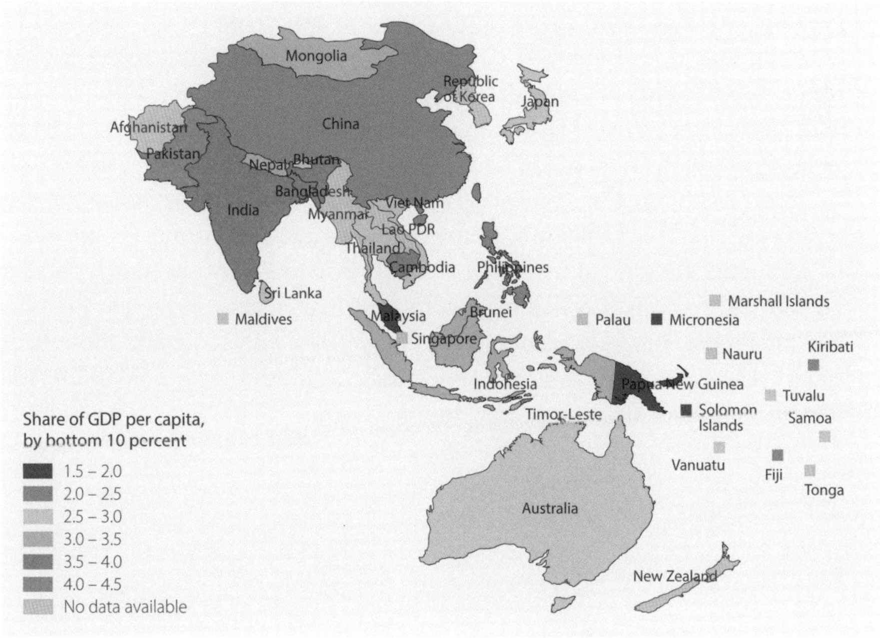

Source of raw data: World Bank (2018)

Figure 2.4 Ratio of female to male literacy rates for adults and youth

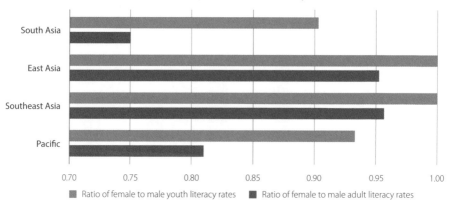

Ratio of female to male youth literacy rates ■ Ratio of female to male adult literacy rates

Source of raw data: World Bank (2018)
Note: Adult refers to those aged 15 and above. Youth refers to those aged 15–24.

Inequality also manifests itself in terms of gender. For example, the gap between male and female literacy among all adults (aged 15 and above) is quite large in some countries in the region, especially in South Asia and the Pacific (the low ratio shown by the red bars in Figure 2.4). Greater gender equality is of course an important aim in its own right, but it also has implications for improved nutrition. Many studies have shown maternal education to be an important determinant of child stunting, even after controlling for the influence of wealth (Cunningham *et al.,* 2017; Headey *et al.,* 2015; Headey and Hoddinott, 2016; Headey, Hoddinott and Park, 2017; Nisbett *et al.,* 2017a, 2017b; O'Donnell, Nicolás and Van Doorslaer, 2007; Raju and D'Souza, 2017; Zanello, Srinivasan and Shankar, 2016), so the education gap represents a serious problem for nutrition outcomes. However, this situation is improving for the young generation, as evidenced by the much smaller literacy gap for youth aged 15–24 years (the higher ratio shown by the blue bars in Figure 2.4). For example, in Bangladesh, young females now have slightly higher literacy rates than young males. Other countries in South Asia however still have large gaps, even among youth.

Beyond literacy, greater female empowerment in terms of control over household financial resources can also have a positive impact on child nutritional status, as dietary diversity tends to increase when women have more say in spending decisions (FAO, 2011a). Higher social status for women is also associated with greater food security for children (Guha-Khasnobis, 2016). Related to this, households where domestic violence is more common are more likely to have undernourished children, so women need to be more empowered in many different ways.

Furthermore, some evidence from South Asia indicates that there is discrimination against women in the intra-household allocation of food (Harris-Fry *et al.*, 2017). As a result, pregnant women in India are far more likely to be underweight than women in sub-Saharan Africa (Coffey, 2015). The low status of women plays a significant role in women's poor nutritional status and in turn children's nutritional status (Mehrotra, 2006; Mukherjee, 2009; Smith *et al.*, 2003). Indeed, one of the reasons South Asia has such high malnutrition rates is because many babies are born small-for-gestational age (SGA), which is largely determined by maternal nutrition status (Christian *et al.*, 2013; Katz *et al.*, 2013).

Women's empowerment varies within the region. East Asia and the Pacific has a very high Gender Development Index (GDI), second only to Latin America and the Caribbean among developing country regions. South Asia, on the other hand, has the lowest GDI value among developing country regions.

Evidence from the Women's Empowerment in Agriculture Index (WEAI) is consistent with the GDI. Among the 13 countries from around the world with WEAI scores (Malapit *et al.*, 2014), the highest score comes from a Southeast Asian country (Cambodia, 0.98) and the lowest score from a South Asian country (Bangladesh, 0.66). The WEAI score for Nepal (0.80) is higher than that for Bangladesh, but well below that for Cambodia (these three countries are the only examples from East, Southeast or South Asia; there are none from the Pacific).

In addition to income and gender, inequality also manifests itself according to location. For example, the prevalence of poverty and stunting is consistently higher in rural areas compared to urban areas. In both East Asia and the Pacific and South Asia, more than 75 percent of the poor live in rural areas (World Bank, 2016a) and the prevalence of stunting is consistently higher in rural compared to urban areas (Figure 2.5). However, stunting is still a serious problem in urban areas; it merits serious policy attention that carefully considers the different contexts of stunting in urban and rural areas. For example, sanitation may be more important in urban areas, especially slums, given the high population densities and the resultant ease with which disease can spread. Furthermore, the share of the poor living in urban areas is rising; in a number of countries in the region the number of poor in urban areas increased between 1990 and 2008 (ADB, 2014).

Figure 2.5 Prevalence of stunting in rural and urban areas

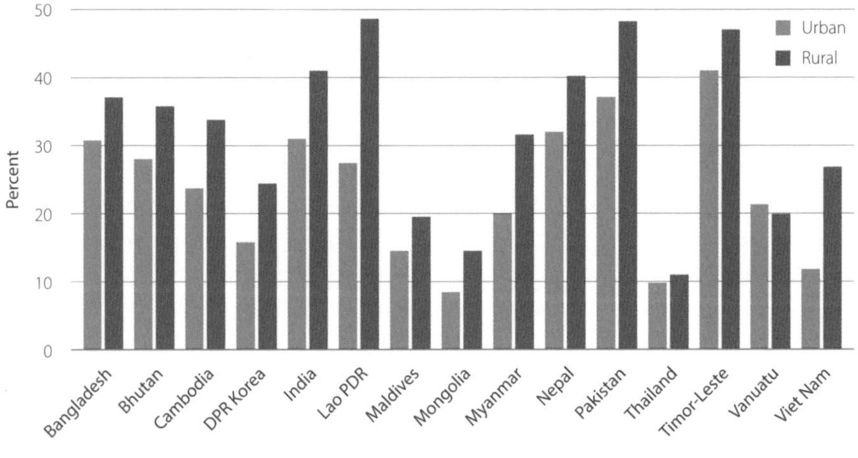

Source of raw data: WHO (2018)

Population growth and demographic shifts

Although population growth is slowing, it is still positive in most countries in the Asia-Pacific region and around the world. The medium projection variant from the United Nations expects that the global population will reach 8.6 billion in 2030 and 9.8 billion in 2050 (Figure 2.6).[2] The continued growth in population has important implications for food demand – FAO (2017) estimates that the world will need to produce 50 percent more food between 2013 and 2050. In this regard, it is not only growth in Asia's population that is important, but growth around the world as well, because population growth elsewhere will affect demand in global food markets, with effects on Asia's import bills and export earnings. Continued population growth (and economic growth) also puts additional pressure on the natural resource base that supports food production – these trends are discussed in Chapter 3.

Economic growth has led to a demographic transition: a transition from high to low rates of fertility and mortality. Declines in fertility and mortality have occurred in all subregions. Projections suggest that birth rates will continue declining until at least 2050; with this decline, the proportion of young people in the population will decline and the proportion of old people will increase. By 2030, the old age dependency ratio will be 45 in regional high-income countries and 25 in East Asia (Figure 2.7). This ratio is positively correlated with GDP per capita: after the high-income countries and East Asia, Southeast Asia comes next, followed by South Asia and the Pacific.

The ageing population affects both rural and urban areas, and has implications for the agriculture sector, as well as for health care systems and safety net programmes. One implication is that the agricultural labour force becomes older, most markedly in developed countries but increasingly also in middle-income countries. Outmigration of young people from rural areas to seek jobs in urban areas further accelerates the ageing of the agricultural labour force. The ageing of the agricultural labour force is described in more detail in Chapter 6, along with its implications for production systems.

---

2 95 percent confidence intervals for the projections are 8.4 to 8.7 billion in 2030 and 9.4 to 10.2 billion in 2050.

Figure 2.6 Population, past and projected, 1950–2050

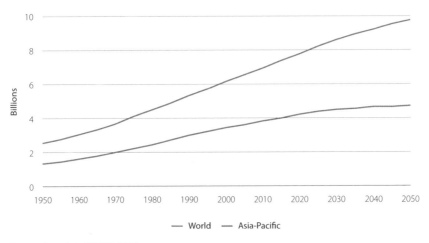

Source of raw data: UNDESA (2017)

Figure 2.7 Old age dependency ratio (ratio of population aged 65+ per 100 population aged 15–64) by subregion, 1950–2050

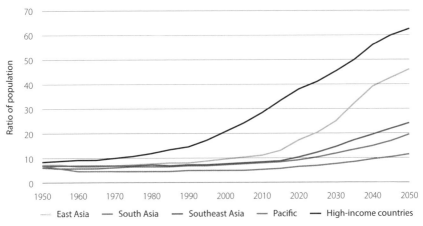

Source of raw data: UNDESA (2015)

Urbanization, migration and remittances

The world is now more than 50 percent urban[3] and soon the Asia-Pacific region will also cross that threshold. Currently, 1.9 billion people in Asia and the Pacific live in urban areas, and this number will increase to 2.5 billion by 2030, representing 50 percent of the world's urban population. Not only does economic growth lead to urbanization, but urbanization also leads to economic growth (Spence, 2009), as cities have lower internal (and often external) costs of transport and generate knowledge spillovers (i.e. people learn more easily from one another when they live and work close together).

Within the region (and throughout the world), urbanization is correlated with income – it is more advanced in the subregions that have higher levels of GDP per capita (Figure 2.8). Looking forward, urbanization is projected to take place in all of the region's countries. Although most of the world's megacities are located in Asia, urban areas are much more than just large national capitals. Urban populations live in a diverse range of megacities, large cities, medium-sized cities and small cities, all of which have different characteristics and interact differently with rural areas (FAO, 2017). The physical expansion of urban areas has implications for the natural resource base that supports agricultural production (see Chapter 3). Urbanization also has implications for diets and obesity (see Chapter 4) as well as implications for value chains, which will need to move more food to urban areas.

Figure 2.8 Urban share of population (current and projected)

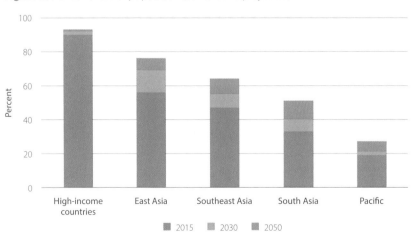

Source of raw data: FAO (2018)

---

3   The definition of urban uses a number of criteria and varies from country to country, often substantially. Roughly speaking, many countries define an urban area as an administrative unit with population being above a certain threshold, often in the range from 2 000 to 5 000 (Deuskar, 2015).

Rural populations have been declining for many years in absolute terms in both high-income countries as well as East Asia (Figure 2.9). In Southeast Asia, the rural population reached its peak in 2013. The situation in South Asia is different, as the rural population is still increasing, but it is predicted to reach a peak around 2030 and will then decline as well. The rural population in the Pacific, however, is still expected to increase until 2050. Thus, rural labour availability will vary substantially across different countries over the next 20 years, which will have key implications for rural wages, labour shortages, mechanization and farm size (see Chapter 6).

Figure 2.9a Rural population, by subregion, 1950–2050

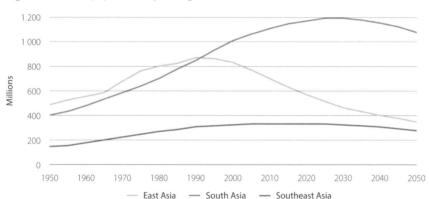

Figure 2.9b Rural population, by subregion, 1950–2050

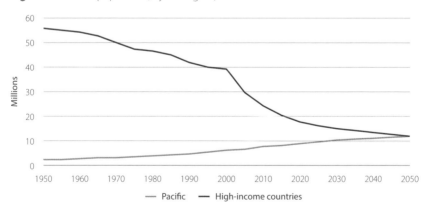

Source of raw data: FAO (2018)

In search of jobs, young people tend to migrate more to urban areas compared to older people. As a result, people of prime working age (defined here as 15–49 years of age), relative to those aged 50 and above, are more likely to live in urban areas. For all countries in the region, high- and low-income alike, the ratio of people of prime working age to the rest of the population is greater in urban than in rural areas (Figure 2.10).[4] In contrast, rural areas have a greater proportion (than urban areas) of both young (14 years and under) and old people (50 years and above).

Many development discussions have noted the outmigration of males seeking employment in urban areas leading to a feminization of rural areas, in turn suggesting a feminization of agriculture (e.g. World Bank, 2016b). The data suggest that, within Asia and the Pacific, this is primarily a South Asian phenomenon.

Figure 2.10 Ratio of number of people of prime working age (15–49 years old) to number of people not of prime working age

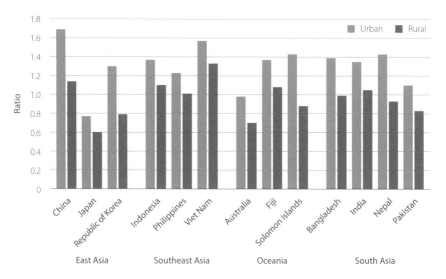

Source of raw data: UNDESA (2017)

---

4   Throughout this publication, Oceania refers to the aggregate of Australia, New Zealand and the Pacific Islands, including all countries for which data are available.

In East Asia, Southeast Asia and the Pacific, among prime working age people (aged 15–49 years), urban areas have a higher percentage of women compared to rural areas in 17 out of 21 countries (Figure 2.11 shows a selection of these countries – the blue bar for urban areas is typically higher than the red bar for rural areas). The only exceptions are China (where the difference is small), Timor-Leste, the Solomon Islands and Vanuatu. By contrast, in South Asia, rural areas have a higher percentage of women compared to urban areas in eight out of nine countries for which there are data (here, the red bar is higher than the blue bar), with the differences being large in Nepal, Bangladesh and Pakistan (and the only exception being the Maldives). The implications of these trends for the gender composition of the agricultural labour force are discussed in Chapter 6.

Permanent migration from rural to urban areas is not the only type of migration. Although data are typically quite limited, rural to rural migration is also common, as is seasonal migration where farm workers move to cities during non-farming seasons to perform casual labour (often infrastructure work) and then move back again during peak harvesting/planting seasons. These movements help to alleviate seasonal labour shortages in agriculture and the remittances generated can also lead to investment in agriculture and nutrition (see Chapter 6).

Figure 2.11 Sex ratio in urban and rural areas for people in the age group 15–49 years old, selected countries

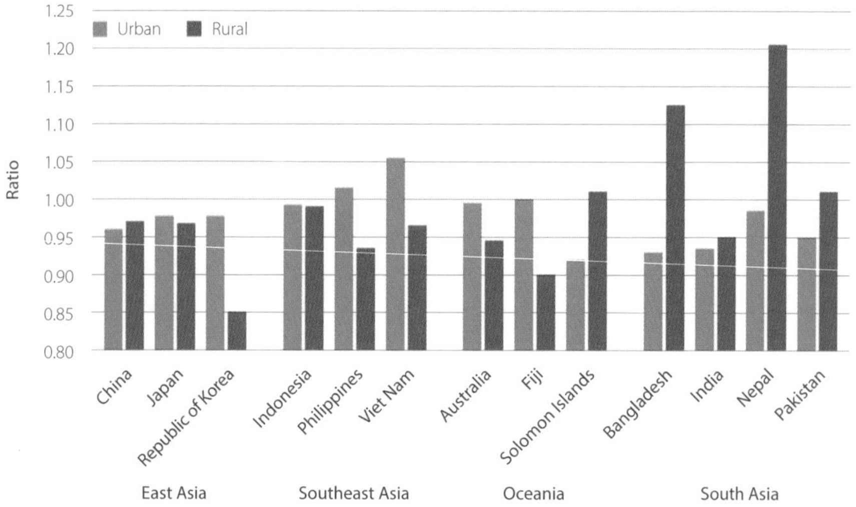

Source of raw data: UNDESA (2017)

Globalization and international trade

As is true for most of the world, the Asia-Pacific region is undergoing economic, social and political globalization (Gygli, Haelg and Sturm, 2018). This involves greater migration, tourism, trade, communication and other activities that bring people from different cultures in contact with each other. This contact influences food preferences and eating habits (see Chapter 4), as well as labour supplies in production systems (see Chapter 6).

Overseas migration is also becoming more important, and very diverse. While many migrants move to high-income countries (either to Organisation for Economic Co-operation and Development [OECD] countries or to the Near East), a quarter of East Asian migrants and a third of South Asian migrants have migrated to another low- or middle-income country. Although there are more male migrants than female, overall the differences are small. Most migrants are lower skilled workers, but over a third of migrants who left for OECD countries have a tertiary education. In total, there are nearly 70 million international migrants from the Asia-Pacific region, slightly less than 2 percent of the total population (World Bank, 2016c).

While India and China receive the largest amount of remittances, countries such as Nepal and many Pacific Islands are far more dependent on remittances (remittances account for 29 percent of GDP in Nepal and in Tonga, Samoa, the Marshall Islands and Tuvalu they range from 11 to 28 percent of GDP). The Philippines, Pakistan, Bangladesh, Indonesia and Viet Nam all ranked in the top 15 of remittance-receiving countries in 2014, with remittances accounting for 10 percent of GDP in the Philippines. Remittances are a significant component of livelihood strategies for many households in the Asia-Pacific region, and networks of migrants can provide key financial and at times technical inputs in food production. However, international migration can also have negative family and social consequences, and is often undertaken because of a lack of government investment in public goods in rural areas. Remittances might also act as a disincentive to work for those who remain behind and become accustomed to receiving money without much effort.

Another particular aspect of globalization involves increased trade in food. The growth of international food trade was relatively slow in the region in the 1990s, but over the past 12 years the value of international food trade, after adjusting for inflation, more than tripled in all three Asian subregions and grew 67 percent in the Pacific (Figure 2.12). Free trade agreements, coupled with investments and innovations in infrastructure that allow for more efficient shipping of goods, have facilitated this increase in trade. The increased trade allows for more efficient food production at the global scale, and provides export opportunities for farmers and value chains, but accordingly also makes it more important for farmers (and value chains) to enhance their competitiveness in order to compete against imports.

Increased competitiveness against imports can be achieved through productivity-enhancing investments in farm production and value chains and/or through higher import tariffs. Tariffs can facilitate the adoption of new technologies and help countries manage price volatility from international markets (Dawe and Timmer, 2012), but there are drawbacks to such an approach, especially if the import barriers are high and persist for a long period of time. Persistently high food prices often harm the poor (FAO, 2011b) and can also compromise nutritional outcomes by negatively affecting the affordability of a healthy diet (see Chapter 4), even when the higher prices are only temporary (Block *et al.*, 2004; D'Souza and Jolliffe, 2010).

Figure 2.12 Annual average value of international trade in food, adjusted for inflation

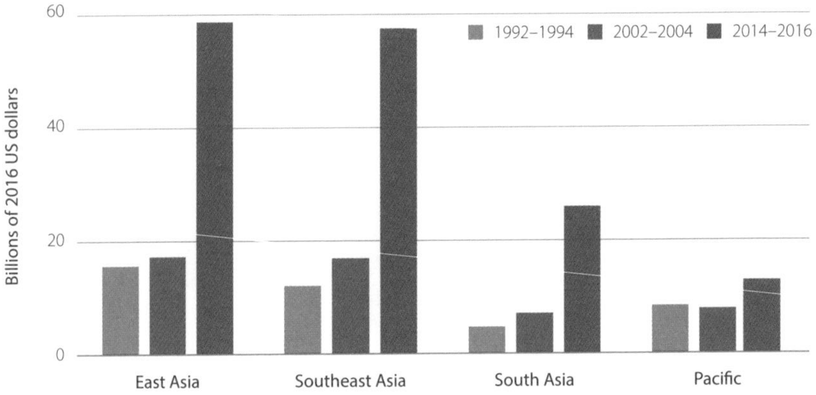

Source of raw data: FAO (2018)

Note: Data for the Pacific are in hundreds of millions of dollars.

Greater trade can provide countries with the chance to import food from regions where land or water are more abundant, but it can also facilitate more rapid exploitation of the environment, especially if exporting countries subsidize the exploitation of natural resources (see Chapter 3).

Despite the rapid growth in international trade, it is still important to note that domestic trade is typically much more important than international trade. Nearly all low- and middle-income countries in the region source more than 85 percent of their dietary energy supply from domestic production, not imports (the main exceptions being several island, peninsular or landlocked nations; Map 2.3). Regardless of the channel, increased trade, both domestic and foreign, requires the development of more sophisticated value chains and more investment in food safety (see Chapter 5).

Map 2.3a Share of dietary energy (calories) produced domestically, Asia

Map 2.3b Share of dietary energy (calories) produced domestically, Oceania

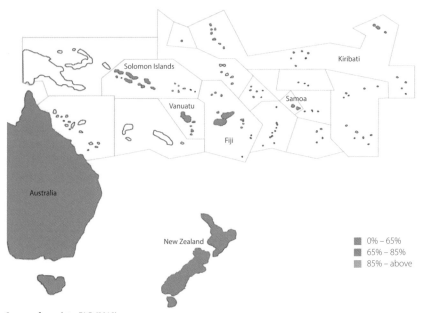

Source of raw data: FAO (2018)

Note: Australia has been cropped in order to show other countries more clearly. Pacific countries without their names shown have no data available.

Structural transformation

In addition to their effects on mortality and fertility rates (which combine to generate the demographic transition) and diets (see Chapter 4), economic growth and urbanization also increase consumer demand for non-food items – consumer electronics (radios, televisions, mobile phones), education, health, entertainment and more convenient modes of transportation (motorcycles, cars), among others. Consumer demand for food is by way of its nature limited to an extent, while demand for other products can be practically boundless, creating more numerous opportunities for employment in the companies that produce those products.

In order to satisfy the increased consumer demand for non-food items, sectors other than agriculture will grow more rapidly (Figure 2.13) and become more important to the economy, while agriculture's share declines (Figure 2.14).[5] However, the reduced relative importance of agriculture is not the result of a shrinking agriculture sector. Agriculture (driven by increased food demand) continues to grow in developing countries – it just grows less rapidly than the industrial and service sectors (Figure 2.14). It should be noted that the agriculture sector is shrinking in some high-income countries (primarily in Japan) but continues to grow in most other high-income countries. Moreover, agricultural growth in the Pacific, while positive, has been very much slower than in East, Southeast and South Asia (see Box 2 in Chapter 4).

Figure 2.13 Average annual growth rates of value-added (GDP) in agriculture, industry and services, 2000–2016

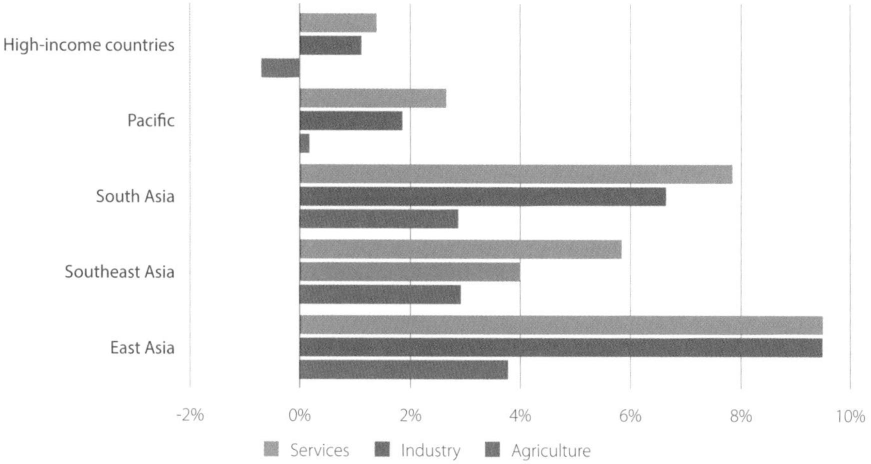

Source of raw data: World Bank (2018)

---

5   In theory, this need only hold true for the world as a whole. It is possible that global consumer demand for non-food products could be satisfied by a small group of countries, with other countries maintaining agriculture as the dominant sector of the economy as they become wealthier. In practice, this has never been observed over the course of economic development, even for countries like the Netherlands and Denmark that have strong agriculture sectors. In other words, changing consumer demand leads to a structural transformation of all economies, namely reduced shares of the agriculture sector in both GDP and the labour force, and increased shares for the industrial and service sectors.

Figure 2.14 Percentage contribution of agriculture to GDP by subregion, 1994–2016

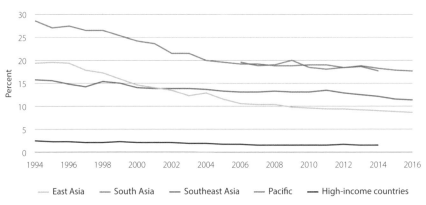

Source of raw data: World Bank (2018)

For low-income countries, the share of agriculture in the labour force is substantially greater than the share of agriculture in GDP for any given subregion at any given point in time (Timmer, 2018). For example, the share of agriculture in employment in South Asia in 2016 was 44 percent, compared to just 18 percent for the share of agriculture in GDP (Figure 2.15). In essence, there are too many agricultural workers chasing too little agricultural value added. This state of affairs reflects an intersectoral productivity gap for agricultural workers – labour productivity (value added per worker) is lower in agriculture than in the rest of the economy.[6]

The intersectoral productivity gap, coupled with fewer work hours available in agriculture during parts of the year due to seasonality, encourages farmers to spend more time in the non-farm economy in urban areas, rural areas or both. The exit of labour from agriculture into the more rapidly growing non-farm economy helps to shrink the intersectoral gap in labour productivity, although the exit may be due to either pull or push forces (Christiaensen, Demery and Kuhl, 2011). The process underlying rising labour productivity

6   In Africa, McCullough (2017) showed that this intersectoral productivity gap is reduced by half when taking account of the fewer hours worked per day in agriculture. Systematic data in this regard are lacking for Asia and the Pacific.

Figure 2.15 Percentage of agriculture in employment by subregion, 1994–2016

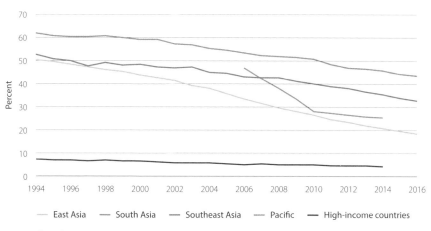

Source of raw data: World Bank (2018)

(value added per worker) in agriculture (e.g. rising wages, mechanization, growth of rural non-farm income) in particular and in rural areas in general is critical to creating a more inclusive society, and will be discussed further in the context of agricultural (and rural) transformation in Chapter 6.

At the margin, moving a worker from agricultural employment to industrial or modern service employment will be a source of economic growth (Timmer, 2018). Not only does this move contribute to economic growth, it can also play a key role in poverty alleviation, women's empowerment and even agricultural development. As one example, non-farm employment in the garment industry in Bangladesh and Cambodia has increased the incomes of the poor (Hossain, 2011; World Bank, 2016a). In addition, because the garment industry employs mostly women, it has increased female empowerment (Hossain, 2011), which can lead to improved nutritional outcomes (although there could also be negative effects if this leads to less time for child care). Finally, urban jobs can be a source of remittances to rural areas, which can be used for investments in agriculture (Mejía-Mantilla and Woldemichael, 2017; Rozelle, Taylor and DeBrauw, 1999, see Chapter 6).

In addition to consumers spending more income on non-food items, a part of the increased income also goes to higher quality food with attributes other than just dietary energy (e.g. taste, convenience, status, nutrition, novelty) – some of these dietary trends are discussed in Chapter 4. Because these demands increase as people get wealthier, the ratio of agribusiness value added relative to agricultural value added tends to increase with economic growth (Figure 2.16). The rising importance of agribusiness relative to agriculture has implications for the value chains that connect production systems with food retail, some of which are discussed in Chapter 5.

Figure 2.16 Ratio of agribusiness value-added to agriculture value-added, 2011

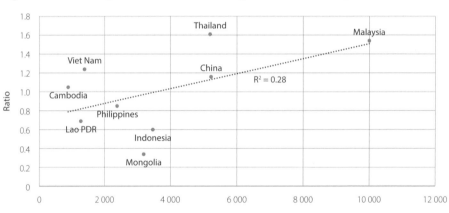

Source of raw data: GTAP (2018)

The growth of agribusiness, such as food and beverage-related manufacturing, also provides employment alternatives to agriculture. Data from countries in the region show that the percentage of manufacturing value added coming from this sector is slightly greater than its percentage of manufacturing employment generated (i.e. most of the points in Figure 2.17 are below the 45° line), suggesting that, on average, food and beverage manufacturing is no better at creating jobs than other forms of manufacturing. Nevertheless, the difference is not large, and food and beverage manufacturing is an important source of employment for those workers who leave agriculture.

Figure 2.17 Share of food and beverages subsector in manufacturing employment and value-added in Asia-Pacific countries

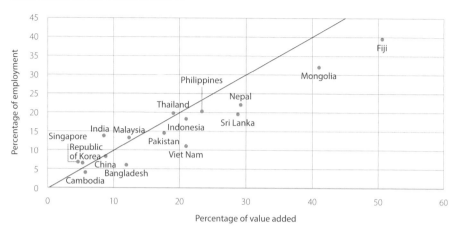

Source of raw data: FAO (2017a)

Note: Line in figure is a 45 degree line.

Summary

The Asia-Pacific region has in general experienced sustained economic growth over several decades. Unfortunately, rising income inequality has accompanied this growth, slowing progress in poverty reduction. In some areas, such as education, there have been reductions in gender inequality, but there is a need for more progress. Economic growth has led to a demographic transition, with reductions in birth rates, slowing (but still positive) population growth and ageing of the population.

Much of the region is also undergoing rapid urbanization due to both rural to urban migration and expansion of urban areas. Economic growth in other parts of the world, coupled with development of and innovations in transport and communications infrastructure, has led to a more globalized world and increased international trade. Accompanying these changes, and driven fundamentally by economic growth, economies in the region have undergone a structural transformation, with the importance of agriculture declining relative to industry and services. Despite its relative decline, the agriculture sector is still growing, and the agribusiness sector is also expanding.

References

Alvaredo, F., Chancel, L., Piketty, T., Saez, E. & Zucman, G. 2018. World inequality report 2018. (also available at https://wir2018.wid.world/).

Asian Development Bank (ADB). 2014. *Urban poverty in Asia.* (also available at http://hdl.handle.net/11540/3310).

Block, S.A., Kiess, L., Webb, P., Kosen, S., Moench-Pfanner, R., Bloem, M. & Timmer, P. 2004. Macro shocks and micro outcomes: child nutrition during Indonesia's crisis. *Economics and Human Biology*, 2: 21–44. https://doi.org/10.1016/j.ehb.2003.12.007

Center for Global Trade Analysis. 2018. *Global Trade Analysis Project (GTAP)* [online]. https://www.gtap.agecon.purdue.edu/

Christiaensen, L., Demery, L. & Kuhl, J. 2011. The (evolving) role of agriculture in poverty reduction - an empirical perspective. *Journal of Development Economics*, 96(2): 239–254. https://doi.org/10.1016/j.jdeveco.2010.10.006

Christian, P., Lee, S.E., Angel, M.D., Adair, L.S., Arifeen, S.E., Ashorn, P., Barros, F.C., Fall, C.H.D., Fawzi, W.W., Hao, W., Hu, G., Humphrey, J.H., Huybregts, L., Joglekar, C. V., Kariuki, S.K., Kolsteren, P., Krishnaveni, G. V., Liu, E., Martorell, R., Osrin, D., Persson, L.A., Ramakrishnan, U., Richter, L., Roberfroid, D., Sania, A., Kuile, F.O.T., Tielsch, J., Victora, C.G., Yajnik, C.S., Yan, H., Zeng, L. & Black, R.E. 2013. Risk of childhood undernutrition related to small-for-gestational age and preterm birth in low- and middle-income countries. *International Journal of Epidemiology*, 42(5): 1340–1355. https://doi.org/10.1093/ije/dyt109

Coffey, D. 2015. Pregancy body mass and weight gain during pregnancy in India and sub-Saharan Africa. *PNAS*, 112(11): 3302–3307. https://doi.org/10.1073/pnas.1416964112

Cunningham, K., Headey, D., Singh, A., Karmacharya, C. & Rana, P.P. 2017. Maternal and child nutrition in Nepal: Examining drivers of progress from the mid-1990s to 2010s. *Global Food Security*, 13: 30–37. https://doi.org/10.1016/j.gfs.2017.02.001

D'Souza, A. & Jolliffe D. 2010. Rising food prices and coping strategies: Household-level evidence from Afghanistan. *The World Bank Policy Research Working Paper No. 5466.* (also available at http://documents.worldbank.org/curated/en/488541467989994718/Rising-food-prices-and-coping-strategies-household-level-evidence-from-Afghanistan).

Dawe, D. & Timmer, P. 2012. Why stable food prices are a good thing: Lessons from stabilizing rice prices in Asia. *Global Food Security*, 1(2): 127–133. https://doi.org/10.1016/j.gfs.2012.09.001

Deuskar, C. 2015. What does "urban" mean? (also available at http://blogs.worldbank.org/sustainablecities/what-does-urban-mean).

FAO. 2011a. *The state of food and agriculture 2010–2011.* (also available at http://www.fao.org/docrep/013/i2050e/i2050e.pdf).

FAO. 2011b. The state of food insecurity in the world: How does international price volatility affect domestic economies and food security? (also available at http://www.fao.org/docrep/014/i2330e/i2330e00.htm).

FAO. 2017. The state of food and agriculture: Leveraging food systems for inclusive rural transformation. (also available at http://www.fao.org/publications/sofa/en/).

FAO. 2018. *FAOSTAT* [online]. www.fao.org/faostat/

Guha-Khasnobis, B. 2016. Women's status and children's food security in Pakistan. *UNU-WIDER Discussion Paper*, 3(38): 45–66. https://doi.org/10.1093/acprof

Gygli, S., Haelg, F. & Sturm, J.-E. 2018. The KOF globalisation index – revisited. *KOF Working Papers.* https://doi.org/10.3929/ethz-b-000238666

Harris-Fry, H., Shrestha, N., Costello, A. & Saville, N.M. 2017. Determinants of intra-household food allocation between adults in South Asia – a systematic review. *International Journal for Equity in Health*, 16(1). https://doi.org/10.1186/s12939-017-0603-1

Headey, D., Hoddinott, J., Ali, D., Tesfaye, R. & Dereje, M. 2015. The other Asian enigma: Explaining the rapid reduction of undernutrition in Bangladesh. *World Development*, 66: 749–761. https://doi.org/10.106/j.worlddev.2014.09.022

Headey, D., Hoddinott, J. & Park, S. 2017. Accounting for nutritional changes in six success stories: A regression-decomposition approach. *Global Food Security*, 13: 12–20. https://doi.org/10.1016/j.gfs.2017.02.003

Headey, D.D. & Hoddinott, J. 2016. Agriculture, nutrition and the green revolution in Bangladesh. *Agricultural Systems*, 149: 122–131. https://doi.org/10.1016/j.agsy.2016.09.001

Hossain, N. 2011. Background paper – Exports, equity and empowerment: The effects of readymade garments manufacturing. *World development report 2012* – Gender equality and development. (also available at http://hdl.handle.net/10986/9100).

Katz, J., Lee, A.C., Kozuki, N., Lawn, J.E., Cousens, S., Blencowe, H., Ezzati, M., Bhutta, Z.A. & Marchant, T. 2013. Mortality risk in preterm and smal-for-gestational-age infants in low-income and middle-income countries: a pooled country analysis. *The Lancet*, 382(9890): 417–425. https://doi.org/10.1016/S0140-6736(13)60993-9

Malapit, H.J., Sproule, K., Kovarik, C., Meinzen-Dick, R., Quisumbing, A., Ramzan, F., Hogue, E. & Alkire, S. 2014. Women's empowerment in agriculture index: Baseline report. *Measuring progress toward empowerment.* (also available at http://www.ifpri.org/publication/measuring-progress-toward-empowerment?).

McCullough, E.B. 2017. Labor productivity and employment gaps in Sub-Saharan Africa. *Food Policy*, 67: 133–152. https://doi.org/10.1016/j.foodpol.2016.09.013

Mehrotra, S. 2006. Child malnutrition and gender discrimination in South Asia. *Economic and Political Weekly*, 41(10): 912–918. https://doi.org/10.2307/4417941

Mejía-Mantilla, C. & Woldemichael, M.T. 2017. To sew or not to sew? Assessing the welfare effects of the garment industry in Cambodia. *The World Bank Policy Research Working Paper No. 8061.* (also available at http://documents.worldbank.org/curated/en/700631494941118323/To-sew-or-not-to-sew-assessing-the-welfare-effects-of-the-garment-industry-in-Cambodia).

Mukherjee, A. 2009. Eight food insecurities faced by women and girl children: four steps that could make a difference, with secial reference to South Asia. Kathmandu, UNESCAP. http://www.un-csam.org/publication/8FoodInsecu.pdf

Nisbett, N., Bold, M. van den, Menon, S.G., Davis, P., Roopnaraine, T., Kampman, H., Kohli, N., Singh, A., Warren, A. & the Stories of Change Study Team. 2017a. Community-level perceptions of drivers of change in nutrition: Evidence from South Asia and sub-Saharan Africa. *Global Food Security*, 13: 74–82. https://doi.org/10.1016/j.gfs.2017.01.006

Nisbett, N., Davis, P., Yosef, S. & Akhtar, N. 2017b. Bangladesh's story of change in nutrition: Strong improvements in basic and underlying determinants with an unfinished agenda for direct community level support. *Global Food Security*, 13: 21–29. https://doi.org/10.1016/j.gfs.2017.01.005

O'Donnell, O., Nicolás, Á.L. & Van Doorslaer, E. 2007. Growing richer and taller: Explaining change in the distribution of child nutritional status during Vietnam's economic boom. *Tinbergen Institute Discussion Paper TI 2007–008/3.* https://doi.org/10.2139/ssrn.957786

Raju, D. & D'Souza, R. 2017. Child undernutrition in Pakistan: What do we know? *World Bank Policy Research Working Paper No.8049*. (also available at http://documents. worldbank.org/curated/en/810811493910657388/Child-undernutrition-in-Pakistan-what-do-we-know).

Rozelle, S., Taylor, J.E. & DeBrauw, A. 1999. Migration, remittances, and agricultural productivity in China. *American Economic Review*, 89(2): 287–291. https://doi.org/10.1257/ aer.89.2.287

Smith, L.C., Ramakrishnan, U., Ndiaye, A., Haddad, L. & Martorell, R. 2003. The importance of women's status for child nutrition in developing countries. *Research Report No.131*. (also available at http://www.ifpri.org/publication/importance-womens-status-child-nutrition-developing-countries).

Spence, M. 2009. Preface. In M. Spence, P.C. Annez & R.M. Buckley, eds. *Urbanization and Growth*, pp. ix–xvi. IBRD/World Bank. (also available at http://hdl.handle.net/10986/2582).

Timmer, P. 2018. State-level structural transformation and poverty reduction in Malaysia: a multi-commodity approach

UNDESA. 2015. World population prospects: The 2015 revision. Key findings and advance tables. Working Paper No. ESA/P/WP.241. New York. (also available at http://www.un.org/ en/development/desa/publications/world-population-prospects-2015-revision.html).

UNDESA. 2017. World population prospects: The 2017 revision, key findings and advance tables. Working Paper No. ESA/P/WP/248. New York. (also available at https://www.un.org/ development/desa/publications/world-population-prospects-the-2017-revision.html).

World Bank. 2016a. Poverty and shared prosperity 2016: Taking on inequality. Washington, DC, World Bank. http://elibrary.worldbank.org/doi/book/10.1596/978-1-4648-0958-3

World Bank. 2016b. World feminization of agriculture in the context of rural transformations: What is the evidence? Working Paper No. ACS20815. Washington, DC. (also available at http://documents.worldbank.org/curated/en/790991487093210959/ Feminization-of-agriculture-in-the-context-of-rural-transformations-what-is-the-evidence).

World Bank. 2016c. *Migration and remittances factbook 2016*. Third edition. Washington, DC, World Bank. (also available at http://hdl.handle.net/10986/2582).

World Bank. 2018. *World Development Indicators* [online]. https://data.worldbank.org/ products/wdi

World Health Organization (WHO). 2018. *Global health observatory data repository* [online]. http://apps.who.int/gho/data/view.main.v100230?lang=en

Zanello, G., Srinivasan, C.S. & Shankar, B. 2016. What explains Cambodia's success in reducing child stunting – 2000–2014? PLoS ONE, 11(9): e0162668. https://doi.org/10.1371/ journal.pone. 0162668

CHAPTER 3

ENVIRONMENTAL DEGRADATION AND CLIMATE CHANGE THREATEN SUSTAINABLE AGRICULTURE

Given continued population and economic growth, there is a need to deliver more food of higher nutritional value to consumers at prices that are affordable by the poor. To some extent, reductions in food loss and waste can help achieve this objective, but as a practical matter, production increases will also be necessary. However, increasing production will not be easy given the fragile state of the natural resource base. This chapter focuses on (i) the increasing demands placed on natural resources by food systems; (ii) the continued degradation of the natural environment in which food production takes place; and (iii) how this degradation and climate change threaten future food security.

While there is a range of environmental problems related to agriculture and food production, this chapter focuses primarily on land, water and climate change. Resource use has exceeded the carrying capacity of our planet in all three of these areas; agriculture has played an important role in that process (Campbell *et al.,* 2017) and the resulting damage endangers our future food production capacity.

As noted in Chapter 1, land is scarcer in Asia and the Pacific, on a per capita basis, than in any other major part of the world. The scarcity of land has affected agricultural development patterns in the region. Production techniques have tended to maximize the productivity of the scarce resource (land) while being less concerned with improving the productivity of the resource that is in surplus (labour in this case). As a result, the value of output per hectare is higher in Asia than in all other major regions with the exception of Europe, while the value of output per worker is lower than in all other major regions with the exception of sub-Saharan Africa (Pardey, 2011; no data for Pacific Island countries are available).

The relatively high level of output per hectare is due partially to an abundant supply of labour, but also due to high cropping intensity and heavy use of water and material inputs (e.g. fertilizers, pesticides). Thus, cropping intensity (the ratio of harvested area to arable land plus permanent crops) is typically much higher in the Asia-Pacific region than in other parts of the world (see the maps in Portmann, Siebert, & Döll, 2010; Ray & Foley, 2013). Asia also uses more inorganic nitrogen (N) and phosphorus (P) fertilizer than any other continent on a per hectare basis, and is in second place for potassium (K) (Figure 3.1). However, in the Pacific, limited access to fertilizer (due to remoteness and the high cost of imports) constrains the opportunities for increased agricultural production and productivity.

Figure 3.1 Inorganic fertilizer use per hectare of land area, 2015

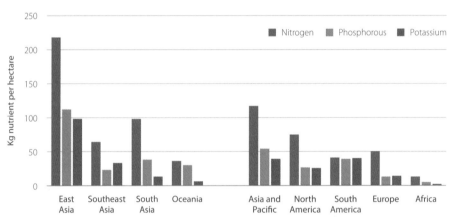

Source of raw data: FAO (2018a). Land area is measured as the sum of arable land and permanent crops.

Note: Oceania refers only to Australia and New Zealand, as no data are available for Pacific countries. Asia-Pacific is a weighted average of the four subregions on the left side of the figure.

While it is possible to intensify production in a sustainable manner (at least in some circumstances), current agricultural production models often damage ecosystems and the services they provide in the process of producing more (and different types of) food. In turn, the loss of these ecosystem services damages our future (and in some cases, current) capacity to produce food.[1] The pressure to rapidly expand food production in response to rapid

---

1   Current production models also damage our environment more broadly than impaired food production capacity – they harm human health through air pollution, use of pesticides, spread of transboundary animal diseases and development of antimicrobial resistance (AMR), harm other species and reduce the amenity value of our natural environment.

economic growth in the region can accelerate this process. Thus, most of the global agricultural systems at risk due to human pressure on land and/or water resources are in the Asia-Pacific region (see Map 3.2 in FAO, 2011).

Globalization has also come into play, with countries 'virtually' importing land and water via food imports that cannot be produced competitively at home due to local natural resource constraints. However, this potential for export earnings may encourage other countries to overexploit their own natural resource base. Some estimates suggest that international trade accounts indirectly for one-third of land use and water withdrawal (Chen *et al.*, 2018). In the current era, "land systems should be understood and modelled as open systems with large flows of goods, people and capital that connect local land use with global-scale factors" (Lambin and Meyfroidt, 2011). The same can be said for water resources as well.

Land area expansion and degradation

Growing demand for land

The economic growth and population growth described in Chapter 2 are leading to increased demand for food. FAO (2017) estimates that global agricultural demand will increase 50 percent from 2013 to 2050. Some of this demand will be met by increased area, but that area expansion will likely be concentrated in Latin America and sub-Saharan Africa. In contrast, the relative scarcity of land in the Asia-Pacific region means that future land expansion will be limited, with most (98 and 95 percent in East and South Asia, respectively)[2] of the increased regional production to 2050 coming from higher yields or increased cropping intensity (Bruinsma, 2011).[3]

Future spikes in global food market prices, like those for many commodities between 2006 and 2011, have the potential to increase land expansion and encourage further granting of large land concessions (Ingalls *et al.*, 2018). These land claims come from both domestic and foreign interests, and while they have the potential to bring additional investment to agriculture and increase food production, they often ignore the rights and interests of the poor and marginalized, including indigenous peoples and ethnic minorities. In the Pacific Islands, most land is under traditional or customary ownership and access to land for agricultural production is a major issue resulting from this type of land tenure system.

---

2 The analysis in Bruinsma (2011) includes Southeast Asia and the Pacific in East Asia.
3 Growing demand for food also puts pressure on marine resources. Global marine fisheries production has increased slightly in recent years, with some of the largest increases coming from within the region, in particular the Western Central Pacific and the Indian Ocean. However, an increasing share of global marine fish stocks is being overfished and this is cause for concern (FAO, 2016).

Rapid rates of urbanization are another source of pressure on land resources. Land cover data show that the area under 'artificial surfaces' nearly tripled between 1992 and 2015, although that category still constitutes just 0.6 percent of the total land area in the region. Looking forward, Bren d'Amour *et al.* (2017) project that Asia will lose more cropland to urbanization by 2030 than any other continent, both in absolute and percentage terms. Within Asia (there are no projections for Pacific countries), Viet Nam, Pakistan and China are projected to suffer the largest percentage losses in production of dietary energy (calories) between 2018 and 2030 due to urbanization (Figure 3.2).

The scarcity of land in the region is reflected in relatively small changes in land planted to temporary or permanent crops in the past two and a half decades, the area of which increased by just a little over 4 percent for the region as a whole between 1990 and 2015. This increase was entirely due to the larger area planted to permanent crops in Southeast Asia, primarily oil palm and rubber, but also cocoa, coconuts and coffee. The land area planted to temporary or permanent crops declined in South Asia and Oceania, and increased only slightly in East Asia.

Figure 3.2 Projected percentage loss in dietary energy (calories) production due to urbanization, 2018–2030

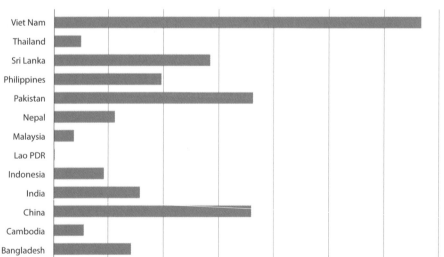

Note: Underlying projections from Bren d'Amour *et al.* (2016) refer to losses between 2000 and 2030. Those estimates were transformed by assuming a constant percentage loss per annum.

In addition to expansion into new areas, production to meet demand can also increase due to higher cropping intensity (more crop area harvested[4] per year on a given piece of land) and higher yields (production per unit area harvested). Among all crops, the largest absolute increase in crop area harvested between 1999–2001 and 2014–2016, for the region as a whole, was for fruit and vegetables, followed by maize, which is used primarily for livestock feed (this is consistent with the findings of Alexander *et al.* [2015] at the global level). Both of these categories are temporary crops. As land area planted to temporary crops decreased in most of the region, increases in 'area harvested' for these annual crops are suggestive of increased cropping intensity. The area under oil crops also increased substantially, with oil palm (a permanent crop) being the major contributor in that category. Aquaculture production has expanded very rapidly, with some percentage being due to area expansion.[5] Consumers are increasingly demanding all of these foods (see Chapter 4), with non-food demands playing a role as well (e.g. palm oil for biodiesel in addition to its use as vegetable oil).

Degradation of land quality and ecosystem services

While (sustainable) intensification will be important to meet future demand, the current manner of intensification too often leads to soil erosion, nutrient mining, topsoil compaction and salinization, as well as loss of biodiversity, potentially damaging future land productivity. Soil erosion also contributes to sedimentation of rivers and irrigation systems, leading to declining water productivity.

There has been much damage to ecosystems throughout the region. Some of the worst situations with respect to soil health in the region occur in highland rainfed cropping systems in the Himalayas. Salinization has occurred extensively in Pakistan, northwest India and northern China due to poor management of irrigation systems and/or poor on-farm water management (FAO, 2011). In parts of China, soils are polluted by heavy metals, which are then taken up in the grain grown in those areas. An estimated 12 million tonnes of polluted grain must be disposed of each year, costing Chinese farmers approximately USD 2.57 billion (Luo *et al.*, 2009). Fruit and vegetable cultivation typically uses more nutrients and pesticides per hectare than staple crops, and increasingly uses plastic mulch, especially in China, which can leave harmful residues in the soil (Cassou, Jaffee and Ru, 2017). Soil degradation and pollution clearly threaten our ability to produce safe food in the future.

---

4   Crop area harvested counts 1 hectare of land planted twice a year as two hectares. Statistics are not available for changes in physical area by crop.
5   FAO does not report cross-country data on the area under aquaculture.

Demand for vegetable oils, coupled with demand for wood products, has led to deforestation in parts of Indonesia and Malaysia, as palm oil is the cheapest major vegetable oil on world markets and is increasingly being used by consumers in Asia and the Pacific (Byerlee, Falcon and Naylor, 2017; Vijay *et al.*, 2016). Notwithstanding a net gain in the region's forest area during the past two decades – primarily due to large-scale afforestation in China – significant areas of natural forests continue to be cleared in many countries. For example, in Asia-Pacific countries reporting decreased forest areas in the period 2000–2015, the sum of net forest area losses totalled 20.3 million hectares (FAO, 2015). Even when afforestation or reforestation takes place, much of it involves the establishment of industrial tree plantations, predominantly for the production of pulpwood, using a limited number of species. While these industrial plantations provide economic and social benefits, only a portion of the ecosystem services and other benefits provided by primary forests are restored. These crucial services include carbon storage, biodiversity and maintenance of wildlife habitats, regulation of water flows, maintenance of soil quality and attenuating erosion (FAO, 2018b). Reduced carbon storage threatens to contribute to global warming, endangering future food production capacity (see the section on climate change in this chapter).

Forest degradation is also leading to a loss of biodiversity and ecosystem services. Degradation results from selective harvesting from forest resources for timber and fuelwood, fire, pests and disease or livestock grazing. While degradation is difficult to detect, recent estimates suggest that it is a significant problem, particularly in tropical forests around the region. Measured in terms of partial canopy cover loss, forest degradation in South and Southeast Asia affected more than 50 million hectares of forested land, more than a quarter of the global total during the period 2000–2012 (FAO, 2015). In Asia and the Pacific, timber production and woodfuel are the main sources of degradation. Degradation is also an under-reported and potentially a significant source of greenhouse gas (GHG) emissions (Pearson *et al.*, 2017).

In sum, while expansion and intensification of agricultural areas contribute to meeting food demand (and other non-food demands as well), too often they have led to land degradation, deterioration of soil quality and loss of biodiversity, all of which compromise our ability to meet food demand in the future. Deforestation and forest degradation reduce the provision of a range of valuable ecosystem services, not only threatening future food production, but also impacting our overall quality of life. Moreover, while the granting of land concessions can bring investment and employment, this also risks ignoring the rights of indigenous peoples, ethnic minorities and other marginalized people, making it difficult for them to escape poverty and achieve a degree of prosperity.

Freshwater scarcity

Growing demand for water and falling water tables

As with land, freshwater is also relatively scarce in the region compared to other parts of the world, at least in parts of the region. Other than the Middle East and North Africa, South Asia and East Asia have the lowest amounts of renewable freshwater resources per capita in the world. Southeast Asia and Oceania, however, have much higher availability of freshwater per capita, and scarcity is primarily a seasonal phenomenon, meaning that water is sufficient to meet demand on an annual basis, but existing and increasing variability results in shortages during the dry season (Figure 3.3).

Figure 3.3 Renewable internal freshwater resources per person, 2014

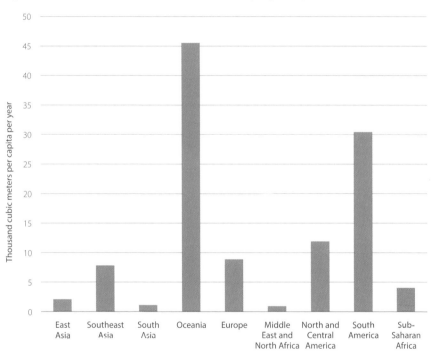

Source of raw data: FAO (2018)

Storage of both surface water and groundwater can be effective for optimizing water availability in time and space for irrigation, leading to increased production and enhanced food security at national and household levels. This is especially true in Asia where a large share (41 percent compared to less than 15 percent on every other continent) of agricultural area is irrigated (Portmann, Siebert and Döll, 2010). The area equipped for irrigation has grown enormously over the past five decades, particularly in China (from 45 to 68 million hectares) and India (from 26 to 67 million hectares) (Scheierling and Treguer, 2016).

Figure 3.4 Percentage distribution of water withdrawals by sector

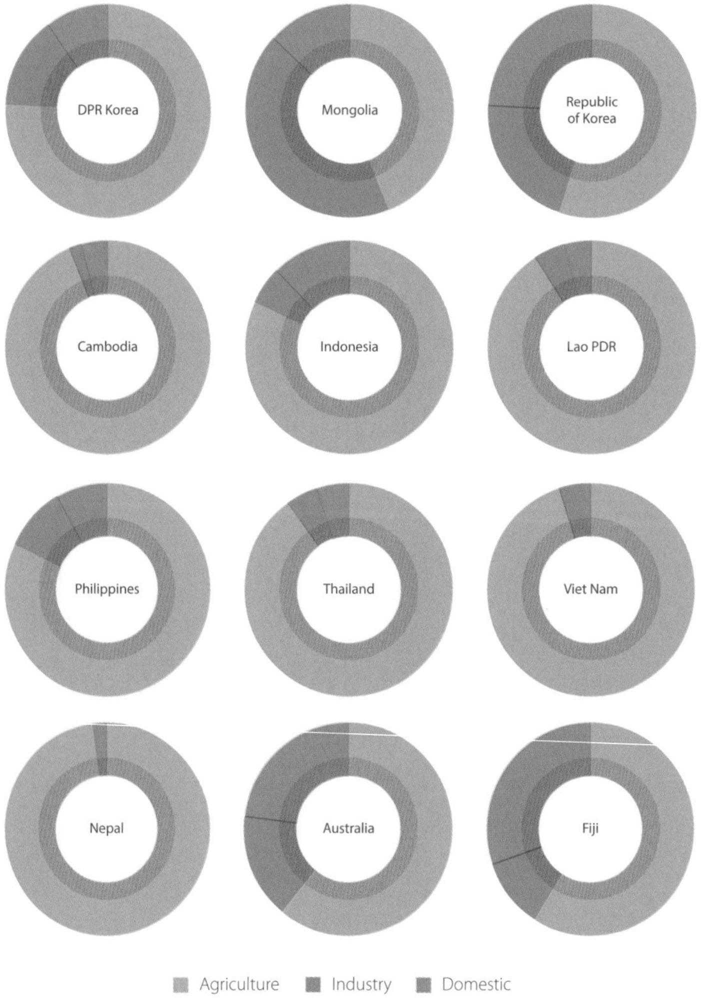

■ Agriculture   ■ Industry   ■ Domestic

Source of raw data: FAO (2018c)

For nearly all countries in the region for which data are available, the bulk of annual freshwater withdrawals is attributable to agriculture (Figure 3.4). In the meantime, agricultural demand for water continues to increase in line with increased food demand. But water demand for domestic and industrial uses tends to increase even faster as countries move from low- to middle-income status. As a result, the share of agriculture in total withdrawals has declined sharply in some upper middle-income countries such as China and Malaysia (Figure 3.5), although the trend is different in South Asian countries as they have a lower level of per capita GDP and thus relatively smaller industrial and domestic demand.

Figure 3.5 Percentage of water allocated to agriculture, selected countries, 1994–2016

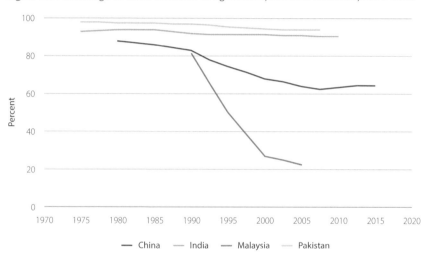

Source of raw data: FAO (2018c)

These multiple demands put pressure on the capacity of farmers to produce food, and, facilitated in some cases by electricity subsidies, encourage greater use of groundwater. While the growing use of groundwater has improved the livelihoods of millions of rural people (Kulkarni, Shah and Vijay Shankar, 2015), it has also caused depletion of aquifers. The situation is perhaps most urgent in India, where six of the most important agricultural states are overexploiting groundwater to meet current irrigation demands and rates of groundwater depletion were the highest in the world from 2002 to 2008 despite above-normal precipitation for the period (Birkenholtz, 2017). According to data from NASA's Gravity Recovery and Climate Experiment (GRACE), rapid depletion has continued into more recent years (Map 3.1). When groundwater tables decline sufficiently, the costs of pumping water from greater depths leads to increased production costs and lower profits for farmers.

Map 3.1 Cumulative total India freshwater losses, 2002–2015

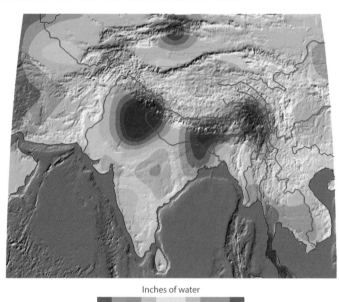

Inches of water

-15 -12 -9 -6 -3 0 3 6 9 12 15

Source: NASA (2015)

Declining water quality

Water quality is deteriorating rapidly throughout the region due to nutrient overloads, industrial pollution and aquifer depletion, leading to concerns over food safety, drinking water quality and food production capacity. Nutrient-use efficiencies are low throughout the region, which means that most applied nutrients are not converted to biomass, but are lost into the environment (Bodirsky *et al.*, 2012; Liu *et al.*, 2010; Yan *et al.*, 2014). Fertilizer subsidies exacerbate this problem by encouraging farmers to over apply nutrients (Gulati and Sharma, 1995; Osorio *et al.*, 2011), which further reduces efficiency and leads to increased discharges into the environment.

There are similar issues with regard to waste from the livestock and aquaculture sectors. While animal manure is a valuable resource for maintaining or improving soil fertility, the geographical concentration of an increasing proportion of the region's modern livestock industry in areas with little or no agricultural land has led to high nutrient overloads in local environments. Untreated manure and faeces-laden waste is often dumped in waterways and on agricultural lands. For example, an estimated 36 percent of livestock waste generated in Viet Nam is dumped directly into the environment without treatment (Cassou, Jaffee and Ru, 2017), resulting in bacterial and microbial overload. Waste from industrial piggery operations is the primary source of nutrient loading affecting waterways in China, Thailand and Viet Nam, contributing from 14 to 72 percent of nitrogen accumulations (Reid *et al.,* 2010).

These nutrient overloads due to expanding livestock production (and excessive fertilizer use) have an impact on drinking water quality through leaching of nitrates and, in the case of livestock waste, possible pathogen transfer. In the Philippines and Thailand, for example, drinking water from 30 percent of all groundwater wells sampled showed nitrate (NO_3) levels above the World Health Organization (WHO) safety limit of 50 mg/litre (Tirado, 2007). Coupled with industrial effluents, they also affect food safety (Box 1). Water quality has also been seriously compromised by depletion of aquifers (e.g. arsenic poisoning of drinking water in Bangladesh [BGS & DPHE, 2001]).

Box 1 Water quality and food safety

Water scarcity is intrinsically linked to water quality, as the pollution of water sources often prohibits different types of uses. An estimated 80-90 percent of all wastewater produced in the Asia and the Pacific region is released untreated (WWAP, 2012), polluting ground and surface water resources. In addition to the damage to ecosystem services, limited or unreliable access to water forces farmers to reuse wastewater in a way that is ad hoc and unmanaged, resulting in serious concerns about food safety. In Bangladesh, for example, industrial effluent, urban waste and agrochemicals are disposed of untreated into open water bodies and rivers. This has resulted in a severe and continuing deterioration of water quality across the country, particularly with regard to levels of arsenic, chromium, cadmium and lead (Ali *et al.,* 2016), leading to compromised food safety and unsafe drinking water.

Compromised food safety has two major impacts. First, farmers receive lower prices for their goods as they are locked out of exporting products via official channels. In addition, consumption of food contaminated with pathogens can hinder nutrient absorption and utilization by the human body, damaging food security.

In addition to food safety and drinking water quality concerns, nutrient overloading also has serious consequences for food production. For example, N and P influx into waterways and oceans promotes very high densities of algae and cyanobacteria in water, also known as algal bloom. As these organisms are generally short-lived, large amounts of dead material accumulate and their decay process drains oxygen from the water, creating dead zones that are uninhabitable for fish and plants, as in the East China Sea at the mouth of the Yangtze River (Breitburg *et al.*, 2018; Li and Daler, 2004). Agriculture is the leading cause of such eutrophication in the Yellow and South China seas (Strokal *et al.*, 2014). Similarly, in 2016 researchers discovered a major dead zone in the Bay of Bengal spanning 60 000 square kilometres (Bristow *et al.*, 2017). Such dead zones can devastate fishing grounds and the livelihoods of those who depend on them.

To summarize, exploitation of water resources has made tremendous contributions in meeting food demand and increasing farm incomes. However, freshwater scarcity is now a serious constraint – current water use for irrigation already depends on unsustainable groundwater depletion in some areas (Wada, Van Beek and Bierkens, 2012). The scarcity of water negatively impacts food production in a variety of ways. Farmers (usually poor, marginalized and/ or at the tail end of irrigation schemes) may lack sufficient water to irrigate crops when needed, leading to reduced yields (and incomes) or complete loss of crops and the capital invested in them (Hussain and Biltonen, 2001). Water scarcity may prevent farmers from flushing salts from the soil, reducing future productivity or requiring the land to be abandoned (Seckler, Barker and Amarasinghe, 1999). Water withdrawals, diversions and land-use changes for irrigation have also impaired the provision of valuable ecosystem goods and services, including flood protection, water purification, biodiversity and maintenance of critical habitats including wetlands and estuaries (Rijsberman, 2004). A number of iconic Asian rivers, including the Indus in South Asia and the Yellow River in China, no longer reach the sea for parts of the year (Postel, 2000). Many rivers have become so depleted that they lose their ability to support productive fisheries (Welcomme *et al.*, 2016) or dilute pollutants (see Box 1). Nutrient overloading is also causing deterioration in water quality that further threatens our future food production capacity.

Greenhouse gas emissions and climate change

In Asia and the Pacific (including high-income countries), agriculture accounted for about 10 percent of total GHG emissions in 2014, down from 21 percent in 1990 (World Resources Institute, 2018).[6] GHG emissions from agriculture increased 27 percent during that time, and have more than doubled since 1961 (Figure 3.6), but emissions from other sectors increased much more rapidly than that as nearly all economies underwent structural transformation, with other sectors of the economy growing more rapidly than agriculture (see Chapter 2). Thus, agriculture's percentage contribution declined.

Agriculture in the Asia-Pacific region contributes to GHG emissions in several ways. Livestock broadly considered (enteric fermentation from ruminant animals, plus animal manure use and management) is by far the largest contributor, followed by rice cultivation (due to emissions of methane and nitrous oxide from flooded fields; this occurs mainly in Asia, not the Pacific Islands) and application of synthetic fertilizers (Figure 3.6). Energy use on farms (e.g. for land preparation, groundwater pumping) is also an important direct contributor to emissions, rivalling the contribution from rice cultivation.[7] Agriculture also indirectly influences emissions through land-use change.

Figure 3.6 Agricultural greenhouse gas emissions by type, Asia and the Pacific 1990–2014

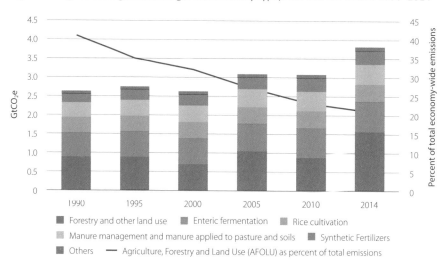

Source of raw data: FAO (2018a)

---

6   Globally, agriculture accounted for 11 percent of total GHG emissions in 2014, similar to the figure for Asia and the Pacific.
7   Energy use in farming is typically included in the energy category, not the Agriculture, Forestry and Land Use (AFOLU) category (according to IPCC guidelines).

Rising temperatures

GHG emissions from both agricultural and non-agricultural activities contribute to increased temperatures and changes in precipitation patterns that affect agricultural production. Temperatures have increased across most of the Asia-Pacific region over the past 60 years, both in terms of average and in terms of extremes. In general, northern latitudes have experienced larger warming trends (Map 3.2). The increase in temperature has been lower in Southeast Asia and the Pacific, although some parts of Malaysia, Mindanao and Sumatra have warmed substantially.

Map 3.2 Linear trends of annual mean temperature, 1951–2012

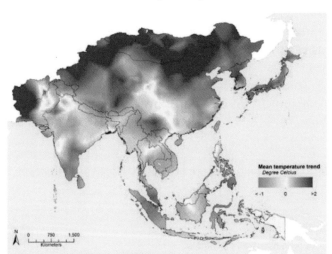

Note: Raw data from Global Climate Monitor system (Camarillo-Naranjo *et al.*, 2018). The trends are derived from Climatic Research Unit (CRU) Time-Series (TS) Version 3.21 of high resolution (0.5 x 0.5 degree) gridded data of month-by-month variation in climate (UEACRU TS3.21, 2013). Map prepared by the International Water Management Institute.

Analyses of the impact of increased temperatures on agricultural production suggest that the effects on production have been noticeable to date, but not massive. Most of this research has focused on a limited number of key staple crops, namely wheat, rice, maize and soybean, because these crops are the main sources of calories and protein for humans, either directly or indirectly through livestock (Lobell and Gourdji, 2012). Climate trends have negatively affected wheat and maize production for many regions of the world, although the effects on rice and soybean yields have been much smaller at the global scale (Lobell, Schlenker and Costa-Roberts, 2011). However, not all effects are negative, with evidence suggesting that warming has benefited crop production in some high-latitude regions, such as Japan and northeast China (Liu, Zhang and Yang, 2016; Zhang *et al.*, 2016). Farmer adaptation can also cushion the impact and some farmers in the region have already been adapting through changes in the cropping calendar, use of stress-tolerant varieties and increased irrigation (Lobell and Gourdji, 2012; Meng *et al.*, 2014, 2016). Looking ahead, it is important to note that the adverse effects of higher temperatures will not necessarily increase linearly with warming. Once temperatures rise sufficiently, threshold effects could lead to larger impacts over a relatively small range of increased temperature (Porter *et al.*, 2014; Schlenker and Roberts, 2009). The threshold phenomenon creates substantial uncertainty in future outcomes that makes planning difficult.

Gradual warming of the atmosphere, and associated changes in the physical and chemical nature of inland waterbodies and the ocean, are also likely to affect marine fisheries and aquaculture. Higher temperatures are affecting the spatial distribution of fish populations and the timing of spawning and migration (Cramer *et al.*, 2014; Johnson *et al.*, 2017). Ocean acidification may affect ocean productivity and capture fisheries, although evidence in this regard is limited (Myers *et al.*, 2017; Rossoll *et al.*, 2012). The evidence base surrounding climate change impacts on livestock is particularly weak, although livestock health is likely to deteriorate under heat stress (Thornton *et al.*, 2009). There may also be impacts on livestock feed intake, feeding efficiency, milk yields and egg laying (Sugiura *et al.*, 2012).

More variable precipitation, increased frequency of extreme events, sea level rise and reduced nutritional content of food

In general, trends in precipitation have been more variable and weaker than those for temperature (Lobell and Gourdji, 2012; Lobell and Burke, 2008; Zhang *et al.*, 2016), although both show increasing and decreasing trends in different parts and seasons of Asia (Map 3.3).

Map 3.3 Linear trends of annual rainfall, 1951–2012

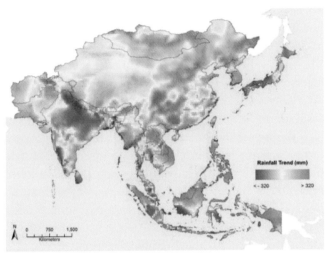

Note: Raw data from Global Climate Monitor system (Camarillo-Naranjo *et al.*, 2018). The trends are derived from Climatic Research Unit (CRU) Time-Series (TS) Version 3.21 of high resolution (0.5 x 0.5 degree) gridded data of month-by-month variation in climate (UEACRU TS3.21, 2013). Map prepared by the International Water Management Institute.

More than trends, however, the key issues for rainfall as it affects agriculture will be timing and variability, which is changing in uncertain ways and differs according to location. Melting glaciers will change river patterns (more water now but then less once they are gone), monsoon winds may shift (with serious implications for planting times) and rainfall may come in shorter, more intense bursts (allowing less time for infiltration into groundwater, increasing erosion). Higher temperatures increase evaporation and hold more water in the air, resulting in a more intensified (extreme) water cycle – lots of water and then long periods with none. The changes in timing are a bigger issue than overall global volumes, especially for agriculture, and have important implications for infrastructure and disaster risk reduction, as well as adaptation in the agriculture sector.

Weather and climate extremes have occurred throughout history but there is evidence of changes in the patterns of climate extremes since 1950 (Alexander, 2016; Cramer *et al.*, 2014; IPCC, 2012; Porter *et al.*, 2014). Further analysis of the findings on record-breaking rainfall events between 1981 and 2010 found large increases in the number of such events in Northern Asia (21 percent), the Tibetan Plateau (31 percent) and Southeast Asia (56 percent). Increased floods and droughts, by increasing the incidence of diarrhoea, can have adverse effects on nutritional deficiencies that lead to stunting and wasting

(Rodriguez-Llanes *et al.*, 2016; Stanke *et al.*, 2013) or increased susceptibility to diseases such as dysentery and malaria that can adversely impact health and nutrition. Increased frequency of extreme climate events obviously requires greater attention to disaster risk reduction and improving agricultural and rural resilience (Hallegate *et al.*, 2015; Hallegate *et al.*, 2016; FAO, 2018d; UNESCAP, ADB and UNDP, 2018).

Over the period 1901–2010, mean sea level rise (SLR) around the globe was 0.19 metres, a larger increase than during the previous 2000 years. This global average masks considerable variability among different parts of the world. Sea level is actually falling in much of the eastern Pacific, while the greatest rise has been in the western Pacific, near the Philippines and among the Pacific Islands (Hoegh-Guldberg *et al.*, 2014).

Gradual SLR will have major effects in the region, including the disappearance of some islands. In addition, SLR will lead to increased coastal flooding and salinization, both of which can affect agricultural systems (including aquaculture).[8] Many of the populations most affected by increased flood risk are in Asia, including Bangladesh, China, India, Indonesia and Viet Nam (Neumann *et al.*, 2015). SLR also affects the distribution of wetlands and mangroves, and the supply of ecosystem services that they provide (Hens *et al.*, 2018).

Climate change is also affecting the nutritional content of food. Recent studies (Ebi and Ziska, 2018; Weyant *et al.*, 2018) indicate that rice, wheat, maize, soybeans, field peas and sorghum grown under high concentrations of carbon dioxide (CO_2) have reduced concentrations of iron and zinc, and often higher concentrations of phytate (which inhibits uptake of various minerals). For poor people who rely heavily on cereals for dietary energy, the reduced micronutrient content of the staple food could have adverse nutritional impacts. There is also emerging evidence that changes in climate could influence the nutritional composition of phytoplankton communities in the marine environment and have wider implications for the nutritional content of ocean catches (Bermúdez *et al.*, 2015).

In addition to their impact on our land and water resources, agricultural practices also contribute to climate change. Climate change, in turn, is affecting food production already and the impacts are likely to increase in the future. Eventually, it may affect diets, food security, nutrition and health in substantial ways (Springmann *et al.*, 2016; Wiebe *et al.*, 2015).

---

8 Groundwater pumping and upstream water consumption is a major cause of land subsidence that exacerbates the impacts of SLR in deltaic and coastal regions (Erban, Gorelick and Zebker, 2014). The trapping of sediments by upstream reservoirs also prevents the natural regeneration of fluvial sediments and increases delta vulnerability to flooding (Syvitski *et al.*, 2009). In most cases, the combination of these upstream anthropocentric activities poses a greater flooding and groundwater salinization risk than that of SLR (Ericson *et al.*, 2006).

Human health

Agricultural practices affect not only environmental health and the sustainability of future food production, but also human health.

Pesticides

Pesticides have been widely used in the Asia-Pacific region for many decades to kill insects, weeds, funguses, molluscs and rodents. They can have beneficial effects on crop production when used appropriately, but also have negative effects on farmers' health when they are applied without proper equipment and training (Pingali and Roger, 1995). If they are sprayed at the wrong time, or in excessive amounts, they can also leave residues on food, affecting the health of consumers. Despite efforts to ban highly toxic pesticides, many markets still sell such products (Cassou, Jaffee and Ru, 2017).

Air pollution

Biomass burning, often to facilitate forest clearing or management of crop residues, is common in the region. The resultant haze and pollution increases airborne particulates that can cause respiratory problems. It obviously affects rural residents, but often it also affects urban dwellers, and can spread across national borders.

Burning of crop residues continues to be practised because it is a cost-effective solution to disposing of those residues in the absence of markets that might purchase them for other purposes. It also can help to manage pests and diseases, and generate increased cropping intensity (thus more profits for farmers) by allowing for the subsequent crop to be planted quickly after harvest of the previous one (Gadde *et al.,* 2009). Burning of such residues is prohibited by law in many countries, but such laws are typically difficult to enforce.

Antibiotic use

Antibiotic drugs are widely used in livestock and aquaculture production systems in order to promote growth and treat disease prophylactically, thus lowering the costs of production per kilogram and making food more affordable. Detailed use data are not available, but sales data suggest that in 2011 the Asia-Pacific region accounted for half the global value of antimicrobials for use in livestock (TMR, 2012). Close to 70 percent of these antimicrobials are used in poultry and pig husbandry.

However, use of antibiotics has serious drawbacks. The more widely the drugs are used, the sooner micro-organisms will develop resistance to them. Because these drugs are often important for treating human disease, the development of bacteria that are resistant to multiple (or all) classes of drugs presents a serious global health care issue. Given the projected growth in demand for animal-source foods, the use of antibiotics has the potential to increase rapidly if measures are not taken to restrict their use.

In Asia, country-level studies on the general trends, impacts and cost of antimicrobial resistance (AMR) are rare. Nevertheless, a recent World Bank report notes that over 45 antibiotics are widely used in Vietnamese livestock and aquaculture production (Cassou, Jaffee and Ru, 2017). Similarly, in China, in a study spanning 1994 to 2000, Zhang et al. (2006) estimated a 22 percent average growth rate of AMR in seven common human bacterial infections, of which three bacteria were also common in livestock. For Thailand, (Prakongsai et al., 2012) reported more than 140 000 drug-resistant infections annually and more than 30 000 patients dying from these infections per year. Overall, comparison of AMR data from various countries reveals that the levels of resistance found across developing Asian countries by far exceed those reported by any of the OECD countries (Chuanchuen et al., 2014).

Animals and diseases

Increased globalization and trade in agricultural products (see Chapter 2) have many benefits, but they can also encourage the spread of animal diseases. When these diseases jump to humans, they have impacts well beyond agriculture that can increase costs substantially. For example, the costs of the SARS epidemic in 2003 were estimated to be USD 54 billion (Lee and McKibbin, 2004) and more than 100 000 people were estimated to have died from the H1N1 swine flu epidemic in 2009 (Simonsen et al., 2013). In fact, over 60 percent of existing and emerging pathogens affecting humans originate in animals (Morens and Fauci, 2013).

In addition to their impacts on human health, these diseases can also have major impacts on the livelihoods of smallholder farmers, who may need to kill diseased animals, thus losing their capital in the process. This additional risk can discourage them from diversifying production beyond basic staple foods. Exposure to animal faeces can also have adverse impacts on human health, with some studies indicating that where poultry are kept at night in family sleeping quarters, there is an increased incidence of child stunting (George et al., 2015; Headey et al., 2017; Penakalapati et al., 2017).

Summary

Agriculture has a wide range of impacts on the environment and human health, and these impacts tend to increase as the sector is asked to produce ever-larger quantities of food to meet the demands of economic and population growth. The impacts on the environment not only affect the quality of the environment, but also affect our ability to produce food sustainably for the future. Agriculture also contributes to global warming and urbanization takes prime agricultural land out of food production. These concerns present major challenges to feed the world with nutritious food at affordable prices in the coming decades.

References

Alexander, L. V. 2016. Global observed long-term changes in temperature and precipitation extremes: A review of progress and limitations in IPCC assessments and beyond. *Weather and Climate Extremes*, 11: 4–16. https://doi.org/10.1016/j.wace.2015.10.007

Alexander, P., Rounsevell, M.D.A., Dislich, C., Dodson, J.R., Engström, K. & Moran, D. 2015. Drivers for global agricultural land use change: The nexus of diet, population, yield and bioenergy. *Global Environmental Change*, 35: 138–147. https://doi.org/10.1016/j.gloenvcha.2015.08.011

Ali, M.M., Ali, M.L., Islam, M.S. & Rahman, M.Z. 2016. Preliminary assessment of heavy metals in water and sediment of Karnaphuli River, Bangladesh. *Environmental Nanotechnology, Monitoring and Management*, 5: 27–35. https://doi.org/10.1016/j.enmm.2016.01.002

Bermúdez, R., Feng, Y., Roleda, M.Y., Tatters, A.O., Hutchins, D.A., Larsen, T., Boyd, P.W., Hurd, C.L., Riebesell, U. & Winder, M. 2015. Long-term conditioning to elevated pCO_2 and warming influences the fatty and amino acid composition of the diatom cylindrotheca fusiformis. *PLoS ONE*, 10(5): e0123945. https://doi.org/10.1371/journal.pone.0123945

Birkenholtz, T. 2017. Assessing India's drip-irrigation boom: efficiency, climate change and groundwater policy. Water International, 42(6): 663–677. https://doi.org/10.1080/02508060.2017.1351910

Bodirsky, B.L., Popp, A., Weindl, I., Dietrich, J.P., Rolinski, S., Scheiffele, L., Schmitz, C. & Lotze-Campen, H. 2012. N2O emissions from the global agricultural nitrogen cycle – current state and future scenarios. *Biogeosciences*, 9(10): 4169–4197. https://doi.org/10.5194/bg-9-4169-2012

Breitburg, D., Levin, L.A., Oschlies, A., Grégoire, M., Chavez, F.P., Conley, D.J., Garçon, V., Gilbert, D., Gutiérrez, D., Isensee, K., Jacinto, G.S., Limburg, K.E., Montes, I., Naqvi, S.W.A., *et al.* 2018. Declining oxygen in the global ocean and coastal waters. *Science*, 359. https://doi.org/10.1126/science.aam7240

Bren d'Amour, C., Reitsma, F., Baiocchi, G., Barthel, S., Güneralp, B., Erb, K.-H., Haberl, H., Creutzig, F. & Seto, K.C. 2017. Future urban land expansion and implications for global croplands. *PNAS*, 114(34): 8939–8944. https://doi.org/10.1073/pnas.1606036114

Bristow, L.A., Callbeck, C.M., Larsen, M., Altabet, M.A., Dekaezemacker, J., Forth, M., Gauns, M., Glud, R.N., Kuypers, M.M.M., Lavik, G., Milucka, J., Naqvi, S.W.A., Pratihary, A., Revsbech, N.P., Thamdrup, B., Treusch, A.H. & Canfield, D.E. 2017. N2 production rates limited by nitrite availability in the Bay of Bengal oxygen minimum zone. *Nature Geoscience*, 10: 24–29. (also available at https://www.nature.com/articles/ngeo2847).

British Geological Survey & The Department of Public Health Engineering (BGS & DPHE). 2001. Arsenic contamination of groundwater in Bangladesh. *British Geological Survey Technical Report WC/00/19*, 1. (also available at http://nora.nerc.ac.uk/id/eprint/11986).

Bruinsma, J. 2011. The resources outlook: by how much do land, water and crop yields need to increase by 2050? In P. Conforti, ed. *Looking Ahead in World Food and Agriculture: Perspectives to 2050*, pp. 233–275. FAO. (also available at http://www.fao.org/docrep/014/i2280e/i2280e.pdf).

Byerlee, D., Falcon, W.P. & Naylor, R.L. 2017. *The tropical oil crop revolution: food, feed, fuel & forests.* New York, Oxford University Press. (also available at https://global.oup.com/academic/product/the-tropical-oil-crop-revolution-9780190222987?).

Campbell, B.M., Beare, D.J., Bennett, E.M., Hall-Spencer, J.M., Ingram, J.S.I., Jaramillo, F., Ortiz, R., Ramankutty, N., Sayer, J.A. & Shindell, D. 2017. Agriculture production as a major driver of the earth system exceeding planetary boundaries. *Ecology and Society*, 22(4). https://doi.org/10.5751/ES-09595-220408

Camarillo-Naranjo JM, Álvarez-Francoso JI, Limones-Rodríguez N, Pita-López MF & Aguilar-Alba M. 2018. The global climate monitor system: from climate data-handling to knowledge dissemination. International Journal of Digital Earth, DOI: 10.1080/17538947.2018.1429502

Cassou, E., Jaffee, S.M. & Ru, J. 2017. *The challenge of agricultural pollution: evidence from China, Vietnam, and the Philippines.* Washington, DC, IBRD/World Bank. (also available at https://elibrary.worldbank.org/doi/abs/10.1596/978-1-4648-1201-9).

Chen, B., Han, M.Y., Peng, K., Zhou, S.L., Shao, L., Wu, X.F., Wei, W.D., Liu, S.Y., Li, Z., Li, J.S. & Chen, G.Q. 2018. Global land-water nexus: Agricultural land and freshwater use embodied in worldwide supply chains. *Science of the Total Environment*, 613–614: 931–943. https://doi.org/10.1016/j.scitotenv.2017.09.138

Chuanchuen R., Pariyotorn N., Siriwattanachai K., Pagdepanichkit S., Srisanga S., Wannaprasat W., Phyo Thu W., Simjee S. & Otte J. 2014. Review of the literature on antimicrobial resistance in zoonotic bacteria from livestock in East, South and Southeast Asia. FAO Regional Office for Asia and the Pacific, Animal Production and Health Commission for Asia and the Pacific.

Cramer, W., Yohe, G.W., Auffhammer, M., Huggel, C., Molau, U., Da Silva Dias, M.A.F., Solow, A., Stone, D.A. & Tibig, L. 2014. Detection and attribution of observed impacts. *Climate change 2014: Impacts, adaptation, and vulnerability. Part B: Regional aspects. Working group II contribution to the IPCC Fifth Assessment Report:* 979–1038. (also available at www.cambridge.org/9781107641655).

Ebi, K.L. & Ziska, L.H. 2018. Increases in atmospheric carbon dioxide: Anticipated negative effects on food quality. *PLoS Medicine, 15(7). https://doi.org/10.1371/journal.pmed.1002600*

Erban, L.E., Gorelick, S.M. & Zebker, H.A. 2014. Groundwater extraction, land subsidence, and sea-level rise in the Mekong Delta, Vietnam. *Environmental Research Letters*, 9(084010). https://doi.org/10.1088/1748-9326/9/8/084010

Ericson, J.P., Vörösmarty, C.J., Dingman, S.L., Ward, L.G. & Meybeck, M. 2006. Effective sea-level rise and deltas: Causes of change and human dimension implications. *Global and Planetary Change*, 50(1–2): 63–82. https://doi.org/10.1016/j.gloplacha.2005.07.004

FAO. 2011. The state of the world's land and water resources for food and agriculture (SOLAW) – Managing systems at risk. (also available at http://www.fao.org/nr/solaw/the-book/en/).

FAO. 2015. *Global forest resources assessment 2015: How are the world's forests changing?* Rome. (also available at http://www.fao.org/forest-resources-assessment/current-assessment/en/).

FAO. 2016. The state of world fisheries and aquaculture: contributing to food security and nutrition for all. Rome. (also available at http://www.fao.org/3/a-i5555e.pdf).

FAO. 2017. *The future of food and agriculture – Trends and challenges.* Rome. (also available at http://www.fao.org/3/a-i6583e.pdf).

FAO. 2018a. *FAOSTAT* [online]. www.fao.org/faostat/

FAO. 2018b. *The state of the world's forests – Forest pathways to sustainable development.* Rome. (also available at http://www.fao.org/state-of-forests/en/).

FAO. 2018c. *AQUASTAT* [online]. http://www.fao.org/nr/water/aquastat/main/index.stm

FAO. 2018d. *The impact of disasters and crises on agriculture and food security 2017.* Rome. (also available at http://www.fao.org/3/I8656EN/i8656en.pdf).

Gadde, B., Bonnet, S., Menke, C. & Garivait, S. 2009. Air pollutant emissions from rice straw open field burning in India, Thailand and the Philippines. *Environmental Pollution*, 157(5): 1554–1558. https://doi.org/10.1016/j.envpol.2009.01.004

Gulati, A. & Sharma, A. 1995. Subsidy syndrome in Indian agriculture. *Economic and Political Weekly*, 30(39): A93–A102. (also available at http://www.jstor.org/stable/4403271).

Hens, L., Thinh, N.A., Hanh, T.H., Cuong, N.S., Lan, T.D., Thanh, N. Van & Le, D.T. 2018. Sea-level rise and resilience in Vietnam and the Asia-Pacific: A synthesis. *Vietnam Journal of Earth Sciences*, 40(2): 126–152. https://doi.org/10.15625/0866-7187/40/2/11107

Hoegh-Guldberg, O., Cai, R., Poloczanska, E.S., Brewer, P.G., Sundby, S., Hilmi, K., Fabry, V.J. & Jung, S. 2014. The ocean. *Climate change 2014: Impacts, adaptation, and vulnerability. Part B: Regional aspects. Working group II contribution to the IPCC Fifth Assessment Report*, pp. 1655–1732. New York. (also available at www.cambridge.org/9781107683860).

Hussain, I. & Biltonen, E. 2001. *Irrigation against rural poverty: An overview of issues and pro-poor intervention strategies in irrigated agriculture in Asia.* Colombo, International Water Management Institute. Proceedings of National Workshops on Pro-Poor Intervention Strategies in Irrigated Agriculture in Asia (9–10 August 2001). (also available at https://hdl.handle.net/10568/38022).

Ingalls, M.L., Diepart, J.-C., Truong, N., Hayward, D., Niel, T., Sem, T., Phomphakdy, M., Bernhard, R., Fogarizzu, S., Epprecht, M., Nanthavong, V., Vo, D.H., Nguyen, D., Nguyen, P.A., Saphanthong, T., Inthavong, C., Hett, C. & Tagliarino, N. 2018. The Mekong state of land. Bern, Centre for Development and Environment & University of Bern and Mekong Region Land Governance. (also available at http://mrlg.org/resources/mekong-state-of-land-brief/).

IPCC. 2012. Managing the risks of extreme events and disasters to advance climate change adaptation. Summary for policymakers. *Special Report of the Intergovernmental Panel on Climate Change.* https://doi.org/10.1017/CBO9781139177245

Johnson, J.E., Bell, J.D., Allain, V., Hanich, Q., Lehodey, P., Moore, B.R., Nicol, S., Pickering, T. & Senina, I. 2017. The Pacific Island region: fisheries, aquaculture and climate change. In B.F. Phillips & M. Pérez-Ramírez, eds. *Climate change impacts on fisheries and aquaculture: A global analysis, I,* pp. 333–379. John Wiley & Sons, Ltd. (also available at https://onlinelibrary.wiley.com/doi/pdf/10.1002/9781119154051.ch11).

Kulkarni, H., Shah, M. & Vijay Shankar, P.S. 2015. Shaping the contours of groundwater governance in India. *Journal of Hydrology: Regional Studies,* 4: 172–192. https://doi.org/10.1016/j.ejrh.2014.11.004

Lambin, E.F. & Meyfroidt, P. 2011. Global land use change, economic globalization, and the looming land scarcity. *PNAS,* 108(9): 3465–3472. https://doi.org/10.1073/pnas.1100480108

Lee, J.-W. & McKibbin, W.J. 2004. *Estimating the global economic costs of SARS.* 92–109 pp.

Li, D. & Daler, D. 2004. Ocean pollution from land based sources: East China Sea, China. *Ambio,* 33(1–2): 107–113. https://doi.org/10.1579/0044-7447-33.1.107

Liu, J., You, L., Amini, M., Obersteiner, M., Herrero, M., Zehnder, A.J.B. & Yang, H. 2010. A high-resolution assessment on global nitrogen flows in cropland. *PNAS,* 107(17): 8035–8040. https://doi.org/10.1073/pnas.0913658107

Liu, Z., Zhang, G. & Yang, P. 2016. Geographical variation of climate change impact on rice yield in the rice-cropping areas of Northeast China during 1980–2008. *Sustainability,* 8(7). https://doi.org/10.3390/su8070670

Lobell, D.B. & Burke, M.B. 2008. Why are agricultural impacts of climate change so uncertain? The importance of temperature relative to precipitation. *Environmental Research Letters,* 3(034007). https://doi.org/10.1088/1748-9326/3/3/034007

Lobell, D.B. & Gourdji, S.M. 2012. The influence of climate change on global crop productivity. *Plant Physiology,* 160: 1686–1697. https://doi.org/10.1104/pp.112.208298

Lobell, D.B., Schlenker, W. & Costa-Roberts, J. 2011. Climate trends and global crop production since 1980. *Science,* 333(6042): 616–620. https://doi.org/10.1126/science.1204531

Luo, Y., Wu, L., Liu, L., Han, C. & Li, Z. 2009. *Heavy metal contamination and remediation in Asian agricultural land.* (also available at http://www.niaes.affrc.go.jp/marco/marco2009/english/program/S-1_LuoYM.pdf).

Meng, Q., Chen, X., Lobell, D.B., Cui, Z., Zhang, Y., Yang, H. & Zhang, F. 2016. Growing sensitivity of maize to water scarcity under climate change. *Scientific Reports,* 6(19605). https://doi.org/10.1038/srep19605

Meng, Q., Hou, P., Lobell, D.B., Wang, H., Cui, Z., Zhang, F. & Chen, X. 2014. The benefits of recent warming for maize production in high latitude China. *Climatic Change,* 122(1–2): 341–349. https://doi.org/10.1007/s10584-013-1009-8

Morens, D.M. & Fauci, A.S. 2013. Emerging infectious diseases: Threats to human health and global stability. *PLoS Pathogens,* 9(7): 7–9. https://doi.org/10.1371/journal.ppat.1003467

Myers, S.S., Smith, M.R., Guth, S., Golden, C.D., Vaitla, B., Mueller, N.D., Dangour, A.D. & Huybers, P. 2017. Climate change and global food systems: Potential impacts on food security and undernutrition. *Annual Review of Public Health,* 38: 259–277. https://doi.org/10.1146/annurev-publhealth-031816-044356

NASA. 2015. Cumulative total India freshwater losses as seen by NASA GRACE, 2002–15. Available at https://images.nasa.gov/details-PIA20206.html.

Neumann, B., Vafeidis, A.T., Zimmermann, J. & Nicholls, R.J. 2015. Future coastal population growth and exposure to sea-level rise and coastal flooding – A global assessment. *PLoS ONE,* 10(6). https://doi.org/10.1371/journal.pone.0131375

Osorio, C.G., Abriningrum, D.E., Armas, E.B. & Firdaus, M. 2011. Who is benefiting from fertilizer subsidies in Indonesia? World Bank Policy Research Working Paper No. 5758. (also available at http://hdl.handle.net/10986/3519).

Pardey, P.G. 2011. African agricultural productivity growth and R&D in a global setting. *Stanford symposium series on global food policy and food security in the 21st century.* (also available at https://fse.fsi.stanford.edu/zh-ch/multimedia/african-agricultural-rd-and-productivity-growth-global-setting-1).

Pearson, T.R.H., Brown, S., Murray, L. & Sidman, G. 2017. Greenhouse gas emissions from tropical forest degradation: an underestimated source. *Carbon Balance and Management,* 12(3). https://doi.org/10.1186/s13021-017-0072-2

Pingali, P.L. & Roger, P.A., eds. 1995. *Impact of pesticides on farmer health and the rice environment.* International Rice Research Institute, Los Baños Philippines and Kluwer Academic Publishers, Norwell, Massachusetts USA. (also available at https://www.springer.com/la/book/9780792395218).

Porter, J.R., Xie, L., Challinor, A.J., Cochrane, K., Howden, S.M., Iqbal, M.M., Lobell, D.B. & Travasso, M.I. 2014. Food security and food production systems. *Climate change 2014: Impacts, adaptation, and vulnerability. Part B: Regional aspects. Working group II contribution to the IPCC Fifth Assessment Report:* 485–533. (also available at www.cambridge.org/9781107641655).

Portmann, F.T., Siebert, S. & Döll, P. 2010. MIRCA2000 — Global monthly irrigated and rainfed crop areas around the year 2000: A new high-resolution data set for agricultural and hydrological modeling. *Global Biogeochemical Cycles,* 24(GB1011). https://doi.org/10.1029/2008GB003435

Postel, S.L. 2000. Entering an era of water scarcity: The challenges ahead. *Ecological Applications,* 10(4): 941–948. https://doi.org/10.2307/2641009

Prakongsai, P. & et al. 2012. Prevention and control of antimicrobial resistance in Thailand

Ray, D.K. & Foley, J.A. 2013. Increasing global crop harvest frequency: Recent trends and future directions. *Environmental Research Letters,* 8(044041). https://doi.org/10.1088/1748-9326/8/4/044041

Reid, R.S., Bedelilan, C., Said, M.Y., Kruska, R.L., Mauricio, R.M., Castel, V., Olson, J. & Thornton, P.K. 2010. Global livestock impacts on biodiversity. In H. Steinfeld, H.A. Mooney, F. Schneider & L.E. Neville, eds. *Livestock in a changing landscape, Volume 1: drivers, consequences and responses,* pp. 111–137. Washington, DC.

Rijsberman, F.R. 2004. Water scarcity: Fact or fiction? *Proceedings of the 4th International Crop Science Congress, 26 Sep – 1 Oct 2004.* (also available at http://www.sciencedirect.com/science/article/pii/S0378377405002854).

Rodriguez-Llanes, J.M., Ranjan-Dash, S., Mukhopadhyay, A. & Guha-Sapir, D. 2016. Flood-exposure is associated with higher prevalence of child undernutrition in rural Eastern India. *International Journal of Environmental Research and Public Health,* 13(210). https://doi.org/10.3390/ijerph13020210

Rossoll, D., Bermúdez, R., Hauss, H., Schulz, K.G., Riebesell, U., Sommer, U. & Winder, M. 2012. Ocean acidification-induced food quality deterioration constrains trophic transfer. *PLoS ONE,* 7(4): e34737. https://doi.org/10.1371/journal.pone.0034737

Scheierling, S.M. & Treguer, D.O. 2016. Enhancing water productivity in irrigated agriculture in the face of water scarcity. *Choices Magazine,* 31(3). (also available at http://www.choicesmagazine.org/choices-magazine/theme-articles/theme-overview-water-scarcity-food-production-and-environmental-sustainabilitycan-policy-make-sense/enhancing-water-productivity-in-irrigated-agriculture-in-the-face-of-water-scarcity).

Schlenker, W. & Roberts, M.J. 2009. Nonlinear temperature effects indicate severe damages to U.S. crop yields under climate change. PNAS, 106(37): 15594–15598. https://doi.org/10.1073/pnas.0906865106

Seckler, D., Barker, R. & Amarasinghe, U. 1999. Water scarcity in the twenty-first century. *International Journal of Water Resources Development*, 15(1–2): 29–42. https://doi.org/10.1080/07900629948916

Simonsen, L., Spreeuwenberg, P., Lustig, R., Taylor, R.J., Fleming, D.M., Kroneman, M., Van Kerkhove, M.D., *et al.* 2013. Global mortality estimates for the 2009 influenza pandemic from the GLaMOR project: A modeling study. *PLoS Medicine*, 10(11). https://doi.org/10.1371/journal.pmed.1001558

Springmann, M., Mason-D'Croz, D., Robinson, S., Garnett, T., Godfray, H.C.J., Gollin, D., Rayner, M., Ballon, P. & Scarborough, P. 2016. Global and regional health impacts of future food production under climate change: a modelling study. *The Lancet*. https://doi.org/10.1016/S0140-6736(15)01156-3

Stanke, C., Kerac, M., Prudhomme, C., Medlock, J. & Murray, V. 2013. Health effects of drought: A systematic review of the evidence. *PLoS Currents*, 5. https://doi.org/10.1371/currents.dis.7a2cee9e980f91ad7697b570bcc4b004

Strokal, M., Yang, H., Zhang, Y., Kroeze, C., Li, L., Luan, S., Wang, H., Yang, S. & Zhang, Y. 2014. Increasing eutrophication in the coastal seas of China from 1970 to 2050. *Marine Pollution Bulletin*, 85: 123–140. https://doi.org/10.1016/j.marpolbul.2014.06.011

Sugiura, T., Sumida, H., Yokoyama, S. & Ono, H. 2012. Overview of recent effects of global warming on agricultural production in Japan. *JARQ*, 46(1): 7–13. https://doi.org/10.6090/jarq.46.7

Syvitski, J.P.M., Kettner, A.J., Overeem, I., Hutton, E.W.H., Hannon, M.T., Brakenridge, G.R., Day, J., Vörösmarty, C., Saito, Y., Giosan, L. & Nicholls, R.J. 2009. Sinking deltas due to human activities. *Nature Geoscience*, 2(10): 681–686. https://doi.org/10.1038/ngeo629

Thornton, P.K., van de Steg, J., Notenbaert, A. & Herrero, M. 2009. The impacts of climate change on livestock and livestock systems in developing countries: A review of what we know and what we need to know. *Agricultural Systems*, 101: 113–127. https://doi.org/10.1016/j.agsy.2009.05.002

Tirado, R. 2007. Nitrates in drinking water in the Philippines and Thailand. *Greenpeace Research Laboratories Technical Note*, 10. (also available at http://www.greenpeace.to/publications/nitrates_philippines_thailand.pdf).

Transparency Market Research (TMR). 2012. Animal feed and feed additives market: Global industry size, market share, trends, analysis and forecast 2011–2018. (also available at https://www.transparencymarketresearch.com/animal-feed-and-feed-additives-market.html).

UNESCAP, ADB & UNDP. 2018. *Transformation towards sustainable and resilient societies in Asia and the Pacific.* (also available at https://www.unescap.org/publications/transformation-towards-sustainable-and-resilient-societies-asia-and-pacific)

United Nations World Water Assessment Programme (WWAP). 2012. *The United Nations World Water Development Report 4: Managing Water under Uncertainty and Risk.* Paris, UNESCO. (also available at http://www.unesco.org/new/en/natural-sciences/environment/water/wwap/wwdr/wwdr4-2012/).

University of East Anglia Climatic Research Unit (UEACRU). [Phil Jones, Ian Harris]. CRU TS3.21: Climatic Research Unit (CRU) Time-Series (TS) Version 3.21 of High Resolution Gridded Data of Month-by-month Variation in Climate (Jan. 1901 - Dec. 2012), [Internet]. NCAS British Atmospheric Data Centre, 2013, Date of citation. Available from http:// badc.nerc.ac.uk/view/badc.nerc.ac.uk__ATOM__ACTIVITY_0c08abfc-f2d5-11e2-a948-00163e251233 ; doi: 10.5285/D0E1585D-3417-485F-87AE-4FCECF10A992

University of Seville. 2017. *Global Climate Monitor* [online]. http://www.globalclimatemonitor.org/

Vijay, V., Pimm, S.L., Jenkins, C.N. & Smith, S.J. 2016. The impacts of oil palm on recent deforestation and biodiversity loss. *PLoS ONE*, 11(7): e0159668. https://doi.org/10.1371/journal.pone.0159668

Wada, Y., Van Beek, L.P.H. & Bierkens, M.F.P. 2012. Nonsustainable groundwater sustaining irrigation: A global assessment. *Water Resources Research*, 48(W00L06). https://doi.org/10.1029/2011WR010562

Welcomme, R.L., Baird, I.G., Dudgeon, D., Halls, A., Lamberts, D. & Mustafa, M.G. 2016. Fisheries of the rivers of Southeast Asia. In J.F. Craig, ed. *Freshwater Fisheries Ecology*. First edition, pp. 363–376. John Wiley & Sons, Ltd. (also available at https://onlinelibrary.wiley.com/doi/abs/10.1002/9781118394380.ch29).

Weyant, C., Brandeau, M.L., Burke, M., Lobell, D.B., Bendavid, E. & Basu, S. 2018. Anticipated burden and mitigation of carbon dioxide-induced nutritional deficiencies and related diseases: a simulation modeling study. *PLoS Med*, 15(7). https://doi.org/10.1371/journal.pmed.1002586

Wiebe, K., Lotze-Campen, H., Sands, R., Tabeau, A., Van Der Mensbrugghe, D., Biewald, A., Bodirsky, B., *et al.* 2015. Climate change impacts on agriculture in 2050 under a range of plausible socioeconomic and emissions scenarios. *Environmental Research Letters*, 10(085010). https://doi.org/10.1088/1748-9326/10/8/085010

World Resources Institute. 2018. *CAIT Climate Data Explorer* [online]. http://cait.wri.org

Yan, X., Ti, C., Vitousek, P., Chen, D., Leip, A., Cai, Z. & Zhu, Z. 2014. Fertilizer nitrogen recovery efficiencies in crop production systems of China with and without consideration of the residual effect of nitrogen. *Environmental Research Letters*, 9(095002). https://doi.org/10.1088/1748-9326/9/9/095002

Zhang, R., Eggleston, K., Rotimi, V. & Zeckhauser, R.J. 2006. Antibiotic resistance as a global threat: Evidence from China, Kuwait and the United States. *Globalization and Health*, 2. https://doi.org/10.1186/1744-8603-2-6

Zhang, Z., Song, X., Tao, F., Zhang, S. & Shi, W. 2016. Climate trends and crop production in China at county scale, 1980 to 2008. *Theoretical and Applied Climatology*, 123(1–2): 291–302. https://doi.org/10.1007/s00704-014-1343-4

CHAPTER 4

DIETS ARE DIVERSIFYING WITH IMPLICATIONS FOR FARMERS AND NUTRITION

Malnutrition in the region: trends and consequences

The Asia-Pacific region is subject to multiple burdens of malnutrition. Lack of sufficient dietary energy and protein, micronutrient deficiencies and overweight and obesity typically co-exist in the same country (Figure 4.1), even within the same household and sometimes within the same individual. In Asia, undernutrition is more prevalent than obesity – currently, obesity rates in Asia are the lowest in the world and are less than 20 percent in all

Figure 4.1 Multiple burdens of malnutrition in the Asia-Pacific region

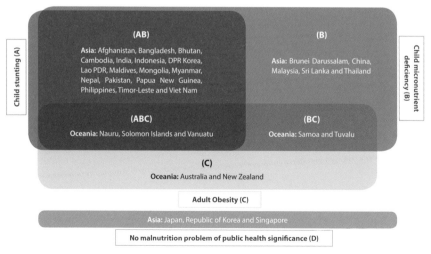

Source of raw data: Modified and updated from FAO (2013)

Note: Countries are categorized based on whether the prevalence of child stunting, child micronutrient deficiency and adult obesity are of 'public health significance.' A problem is considered to be of public health significance if prevalence exceeds 20 percent.

Asian countries. On the other hand, obesity rates in Asia are rising, and it is best to address the problem now, before it becomes more serious. In the Pacific, both undernutrition and obesity already are problems of public health significance. Furthermore, it is important to realize that malnutrition is not only a public health issue, but also has important economic consequences, some of which are highlighted below.

Undernutrition

Broadly speaking, undernutrition is defined as intake of dietary energy, high quality protein or micronutrients in insufficient amounts for a healthy life. This condition can be caused by insufficient food intake, of course, but it can also be due to inadequate utilization of food when people do not have access to safe drinking water and proper sanitation or face a high burden of disease. Poor nutrition, particularly during the critical first 1 000 days of a child's development, causes developmental delays (both physical and cognitive), illness and long-term complications that increase health care costs, reduce educational attainment and lower participation in the workforce, ultimately resulting in reduced individual income-generating capacity (Hoddinott *et al.*, 2013; Hoddinott, 2013). As a result, undernutrition imposes a range of ongoing costs to both households and the national economy.

In the SDG framework, undernutrition is measured using various indicators: the prevalence of stunting, the prevalence of undernourishment, the prevalence of moderate or severe food insecurity measured using the food insecurity experience scale (FIES) and the prevalence of wasting. Most research on the human development consequences of undernutrition has focused on stunting in children under the age of five years (defined as a child being more than two standard deviations below the median height of the distribution), because it measures chronic malnutrition and incorporates both quantity and quality of diet. There is also research that indicates that stunting in the first two years of life can have lifelong consequences (Black *et al.*, 2013; Victora *et al.*, 2008). Because of these lifelong consequences, this chapter will focus largely on stunting as a key indicator.

Figure 4.2 Trends in prevalence of stunting in children under five years of age, by subregion, 1990–2016

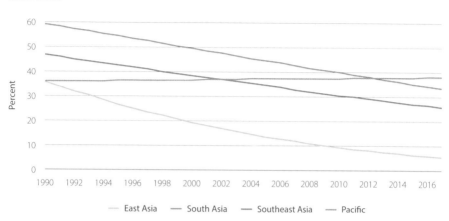

Source of raw data: UNICEF, WHO and The World Bank Group (2018)

Both East Asia and Southeast Asia reached the MDG target of halving the proportion of undernourished (those with insufficient dietary energy) between 1990 and 2015, and South Asia made important progress (FAO, 2016). In terms of stunting, there have also been substantial reductions over the past few decades (Figure 4.2). The Pacific Islands have been an exception to the general improvements, possibly due to the lack of economic growth that affects household income, and the lack of agricultural growth that, coupled with geographic isolation, restricts the availability and accessibility of a diverse diet (see Box 2). The prevalence of stunting in the Pacific is now higher than in any of the other subregions.

Despite the sustained progress in much of the region, the prevalence of stunting remains at a high level in many countries, especially those with low per capita GDP. Fourteen different countries, many of them with large populations, and from all the different subregions except East Asia, have a prevalence of stunting above 30 percent. The prevalence of stunting is higher in rural areas, but it is still very high in urban areas and constitutes a serious problem in both locations (Figure 2.5). In fact, stunting prevalence in the poorest quintile in urban areas is often slightly higher than in the poorest quintile in rural areas (FAO, 2018a). In urban areas, the poorest quintile often lives in crowded slum conditions where the lack of good sanitation is more dangerous than in rural areas. The urban poor may also be more vulnerable to macroeconomic shocks due to the relative lack of capacity to produce their own food (Figure 5.4).

Figure 4.3 Micronutrient deficiency rates among children and women by subregion (latest year available)

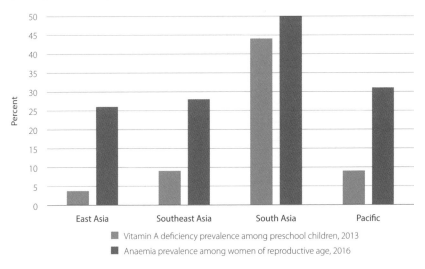

Vitamin A deficiency prevalence among preschool children, 2013

Anaemia prevalence among women of reproductive age, 2016

Source of raw data: GNR (2013) and WHO (2016)

Note: Vitamin A deficiency is estimated for children under the age of five. Anaemia prevalence is estimated for women of reproductive age.

Micronutrient deficiencies are widespread in the region (Figure 4.3), with the most important being iron, vitamin A, iodine and zinc. Lack of micronutrients is often referred to as 'hidden hunger' because while these nutrients are important for good health, people with these deficiencies do not feel hungry, as they do when intake of dietary energy is insufficient. Hidden hunger affects billions of people and leads to stunted growth, disease and poor health conditions. Vitamin A, for example, is an essential nutrient needed by humans for the normal functioning of vision, growth and development and even mild vitamin A deficiency can increase the risk of morbidity and mortality. Between 1991 and 2013, there has been rapid progress in reducing vitamin A deficiency for all subregions other than South Asia, where the incidence remains high (Stevens *et al.*, 2015). Iron deficiency or anaemia during pregnancy increases the risk of a child being born with low-birth weight, as well as increasing the risk of mortality for both mother and child (Young and Ramakrishnan, 2017). A lack of iodine can cause brain damage, compromise physical growth and cognitive development, increase infant mortality, and lead to goiter and hyperthyroidism. Iodine content in most foods is low because many soils in the region are depleted of iodine (de Pee, 2017). As a result, many countries in the region have implemented policies for mandatory fortification of salt with iodine. Data on zinc deficiency are poor, but it is estimated to affect more than 30 percent of the population in several large countries in the region, including Bangladesh, Pakistan, the Philippines and Viet Nam (Hess, 2017). Zinc deficiency contributes to poorly functioning immune systems and higher rates of illness (Micronutrient Initiative, 2015). Micronutrient deficiencies are generally more prevalent in lower-income subregions (Figure 4.3).

Overnutrition

Overnutrition refers to intake of too much energy from food, including low quality food that is high in salt, sugar and saturated fats, relative to the amount of energy expenditure (e.g. exercise, walking, physical labour). Overnutrition has increasingly become a public health, economic, and political problem of utmost importance. Diet-related non-communicable diseases (NCDs, e.g. diabetes, heart disease) are now the leading cause of death and morbidity in many developing countries and are causing escalating health care treatment costs that strain the limited financial capacity of their health care systems (Allotey, Davey and Reidpath, 2014; IFPRI, 2015, 2016; NCD-RisC, 2016). In China, economic losses due to obesity are expected to rise to 9 percent of GDP in 2025 from 4 percent in 2000 (IFPRI, 2016). In Indonesia, health care and productivity losses from obesity are estimated at between USD 2 to 4 billion (Helble and Francisco, 2017). Due to the high financial burden of obesity-related diseases and a lack of publicly funded health care systems in many low- and middle-income countries, a great share of the cost of NCDs falls on households and can lead to poverty (Allotey, Davey and Reidpath, 2014; IFPRI, 2016).

The most commonly used indicator to measure overnutrition is body mass index (BMI), which is body mass (in kilograms) divided by the square of height (in metres). A value between 25 and 30 is considered overweight, while a value in excess of 30 is considered obese. Current levels of obesity in the different subregions vary widely. Obesity rates in the Pacific are the highest in the world, especially in Micronesia and Polynesia – indeed, the countries with the ten highest obesity rates in the world are all Pacific Island countries (NCD-RisC, 2018). In contrast, at present, adult obesity rates in Asia are generally among the lowest in the world, particularly for the high-income countries (Japan, Republic of Korea and Singapore). But they are rising rapidly in all countries (Figure 4.4), consistent with rising sales of unhealthy foods in the middle-income countries (see more in this chapter). In addition, there is evidence that the negative health effects of overweight/obesity begin at a lower value of BMI in Asian populations (Ma and Chan, 2013; Wen *et al.*, 2009; WHO expert consultation, 2004).

In all subregions, women tend to have higher rates of obesity than men. The reasons are not necessarily clear, but may be due to genetics, a higher probability of staying at home due to gender inequality issues and eating relatively more carbohydrates and less protein than men (Bhurosy and Jeewon, 2014).

As a broad generalization, across the entire globe, obesity is more prevalent in higher income countries and in urban areas (relative to rural areas; Figure 4.5). Greater income can lead to additional food consumption due to improved economic access and urban employment tends to be more sedentary than rural work. The relationship between income, urbanization and obesity is however, far from linear. For example, obesity is low in Japan and the Republic of Korea, two high-income countries, as noted above. And within many developed countries, obesity rates are lowest for the wealthiest citizens (Monteiro *et al.*, 2004). Thus, greater income does not lead inevitably to obesity – the importance of diets and urban food environments is discussed in more detail later in the chapter.

Figure 4.4 Prevalence of obesity in adults by subregion, 1975–2014

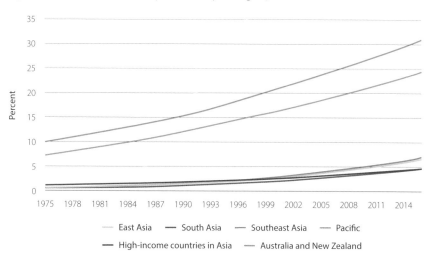

Source of raw data: NCD-RisC (2018)

Note: High-income countries in Asia include Brunei Darussalam, Japan, Republic of Korea and Singapore.

Figure 4.5 Obesity prevalence among adult women, urban and rural areas

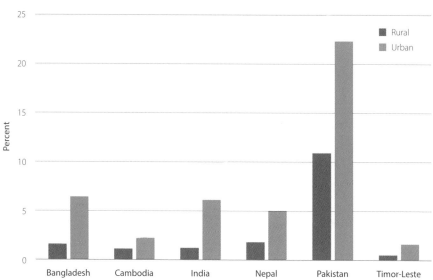

Source of raw data: WHO Global Health Observatory (Bangladesh 2011; Cambodia 2010; India 2005; Nepal 2011; Pakistan 2012; Timor-Leste 2009)

The nutrition transition

The above trends (declining undernutrition where economic growth is strong, rising overnutrition in many countries) are usefully viewed within the context of the nutrition transition, a broad historical pattern that characterizes human activities and diets (Popkin, 2006). These stages include: (i) collecting food; (ii) the rise of settled agriculture and increased risk of famine; (iii) industrialization and more diverse diets with receding risk of famine; (iv) increased incidence of NCDs; and (v) healthier diets and lifestyles (Popkin, 2006). This transition is occurring rapidly in some countries in the Asia-Pacific region due to rapid economic growth and urbanization (Chapter 2). However, not all steps of the transition are inevitable. For example, some developed countries in the region may have skipped the step of increased incidence of NCDs (e.g. Japan, Republic of Korea). The different stages of the nutrition transition co-exist, with different countries and different people experiencing different stages at the same time. The next section of this chapter will discuss trends in dietary structure in the region that underlie the nutrition transition.

Trends in dietary structure

Food consumption patterns have changed substantially in the region over the past few decades. Three particular trends stand out: declining consumption of starchy staples in general and rice in particular; increasing consumption of animal-source foods (ASF) and fruits and vegetables (FV); and increasing consumption of ultra-processed foods that tend to be high in salt, sugar and saturated fats.

Staple foods

Staple foods form the agricultural and cultural foundation of human societies around the globe. In the Asia-Pacific region, rice is the key staple for the largest number of people, but wheat is the staple in some areas (Afghanistan, Pakistan, parts of China and India, Australia and New Zealand), while root crops are the traditional staple in much of the Pacific. In all of these countries, the staple is the most important single source of calories, and in several countries the staple still accounts for half or more of total dietary energy.

After increasing in all subregions for at least parts of the past century, per capita consumption of staple foods has more recently started to level off (in the Pacific) or decline slightly (in East Asia and South Asia) (Figure 4.6). Per capita consumption in Southeast Asia still appears to be increasing, but this may be due to data problems in some countries (Bernadette, Munthe and Taylor, 2016; SKRI, 2016).

Figure 4.6 Apparent consumption of cereals and starchy roots by subregion, 1990–2013

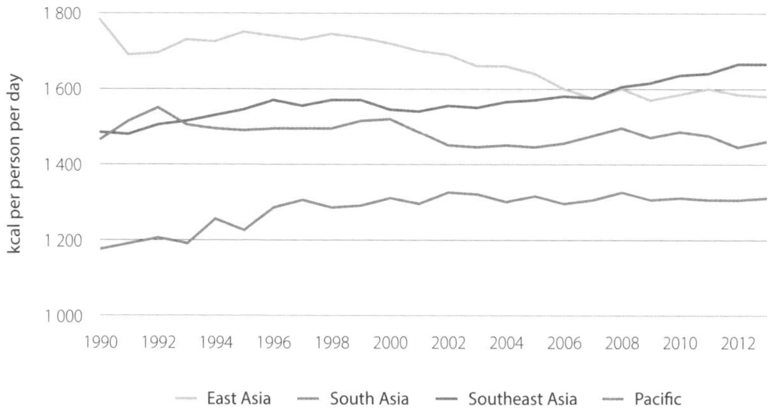

Source of raw data: FAO (2018b)

Note: Data are population-weighted averages. Asian subregions include high-income countries, but Pacific excludes Australia and New Zealand.

Animal-source foods, fruits and vegetables, legumes

Dietary diversity beyond staples is important for nutrition because different foods contain different macronutrients and micronutrients. Staple foods are a good source of dietary energy, but usually lack the key nutrients necessary to prevent disease and various micronutrient deficiencies. As a result, a diet overly reliant on staples is more likely to lead to stunted physical and cognitive growth. The main food groups that contain micronutrients are animal-source foods (including eggs, fish, dairy products), fruits, vegetables and legumes. Animal-source foods in particular are extremely rich in vitamins and minerals such as iron, zinc and vitamin B12 that are either absent or hard to absorb from plant source foods, and are also good sources of high-quality protein. Animal-source foods are extremely important for young children that can eat only limited amounts of food (due to the size of their stomachs), but are often hard to access as they are more expensive and can spoil easily. It is also important to note that not all animal-source foods contain the same nutrients – thus, diversity within food groups is also important. For example, although all animal-source foods provide a good amount of the amino acids contained in high quality protein, essential fatty acids are most easily attainable from fish while minerals such as iron have a high bioavailability in red meat (Allen, 2008; de Pee, 2017).

Plant-source foods can be rich in fibres that help to prevent obesity (Hawkes *et al.*, 2015), and WHO estimates that 5 million deaths worldwide in 2013 were due to inadequate consumption of fruits and vegetables (WHO, 2018). However, some of the micronutrients they possess can have low bioavailability, meaning that the human body does not absorb them readily. Fruits and vegetables are good sources of vitamin C but iron, zinc and vitamin B12 are typically less bioavailable than in animal-source foods. Pulses and legumes are good sources of protein but by themselves cannot provide sufficient amounts of iron, zinc and vitamin B12 for child growth. Thus a variety of foods is required to ensure adequate intake of all essential amino acids.

Consumption of animal-source foods (Figure 4.7a) and fruits and vegetables (Figure 4.7b) has been rising in the region. For animal-source foods, the increases have been most rapid in East and Southeast Asia, with smaller increases in South Asia and the Pacific. The trends for fruits and vegetables were similar, with East Asia experiencing the most rapid growth and the Pacific minimal growth.

Consumption of pulses (Figure 4.7c) has followed a different trend, however. It has been stagnant in East and Southeast Asia, with small increases in the Pacific. South Asia has seen a rise in the past few years, but the level remains much below that in the 1960s, when it was 180 kcal per person per day.

Figure 4.7a Apparent consumption of meat, milk, eggs, fish and seafood by subregion, 1990–2013

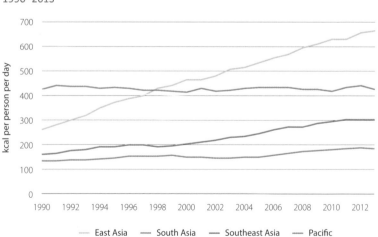

Figure 4.7b Apparent consumption of fruits and vegetables by subregion, 1990–2013

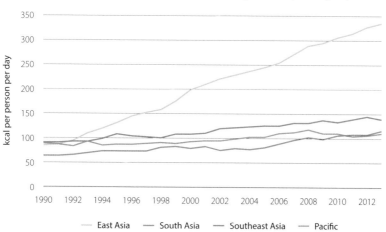

Figure 4.7c Apparent consumption of pulses by subregion, 1990–2013

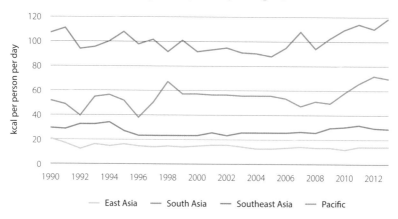

Source of raw data: FAO (2018b)

Note: Apparent consumption is from food balance sheets and is calculated as production plus imports minus exports minus stock accumulations minus uses for industry, seed, feed and waste. Seafood refers to crustaceans, cephalopods and molluscs, but not aquatic plants.

Data are population-weighted averages. Asian subregions include high-income countries, but Pacific excludes Australia and New Zealand.

Fats, sugar and salt

Some amount of fats, sugar and salt are necessary in all diets, but excessive consumption can lead to serious health problems, and many people are now consuming more than is optimal for health. Consumption of vegetable oils and sugar has increased in all subregions since 1990, with Southeast Asia experiencing the largest increases (45 percent in the past 15 years) and now consuming the largest amount of vegetable oils per capita (Figure 4.8a and Figure 4.8b). Data on sales of 'ultra-processed food products and oils and fats' suggest that consumption of such foods is high (and rising) in Southeast Asia relative to East Asia (Baker and Friel, 2016). Both of these observations are consistent with the fact that the obesity rate in Southeast Asia is slightly higher than for East Asia (see Figure 4.4), even though East Asia has higher per capita income. Also of note is that, among high-income countries, Japan, the Republic of Korea and Singapore buy much less of these types of foods than Australia and New Zealand (Baker and Friel, 2016). Thus, a nutrition transition to foods high in fats, sugar and salt with rising incomes is not inevitable – it appears to be connected to the diets people eat, in addition to factors such as physical exercise that are not discussed here.

Figure 4.8a Apparent consumption of vegetable oils and animal fats by subregion, 1990–2013

Figure 4.8b Apparent consumption of sugar and sweeteners by subregion, 1990–2013

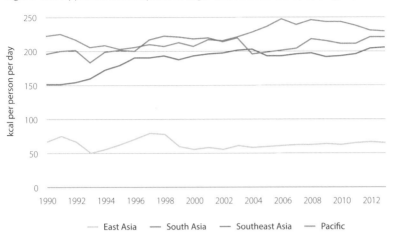

Source of raw data: FAO (2018b)

Note: Apparent consumption is from food balance sheets and is calculated as production plus imports minus exports, minus stock accumulations, minus uses for industry, seed, feed and waste. Sugar and sweeteners from imported processed foods are not included in the above, nor is sugar naturally present in food (e.g. from fruits and fruit juices).

Data are population-weighted averages. Asian subregions include high-income countries, but Pacific excludes Australia and New Zealand.

What is driving the trends in diets and malnutrition?

Several factors are driving the dietary shifts described above. Some of the most important factors affecting consumer demand are growing incomes, changing food prices and urbanization. In addition, changes in the structure of domestic food production are essential in meeting this demand, given that most dietary energy is produced domestically (Map 2.3a and Map 2.3b).

Growing incomes

Despite the importance of staple foods in all of the societies noted above, humans around the world also desire diverse diets. This is reflected in Bennett's Law, which states that starchy staples (rice, wheat, potatoes, millet) account for a smaller share of dietary energy as people get more income, reflecting a desire for dietary diversity (Timmer, 2018). This leads to decreased consumption of these staples on a per capita basis given the limits to energy intake for individuals. Declines in rice consumption have been greater in urban areas than in rural areas, and declines have been greater for those at the upper end of the income distribution compared to the poor (Figure 4.9).

Figure 4.9 Annualised percentage change in rice consumption by quintile and location, Indonesia, India, and Bangladesh

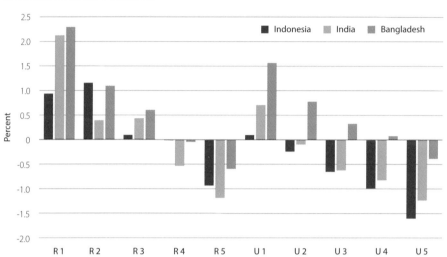

Source of raw data: Timmer *et al.* (2010)

Note: R refers to rural quintiles, U to urban quintiles. Quintile 1 has the lowest income, Q5 the highest. Periods over which changes are calculated are 1967–2006 for Indonesia, 1983–2005 for India, and 1983–2005 for Bangladesh.

Diversification within the category of starchy staples also takes place, again reflecting the desire for diversity, especially for urban residents and young people. For example, urban residents in southern China, where rice is the traditional staple, tend to increase wheat consumption (up to a point) and reduce rice consumption as income increases. The reverse is true in urban areas in northern China, where traditional wheat eaters tend to increase rice consumption (again, up to a point) but reduce wheat consumption as income increases (Timmer, Block and Dawe, 2010). Malaysian families with a younger head of household tend to eat more wheat than families with an older head of household, even after controlling for income (Timmer, Block and Dawe, 2010).

Data from household surveys show that income is a key driver for consumption of several different types of nutritious foods (as well as foods that are less nutritious, see below). For example, consumption of vegetables, meat and fish, eggs, milk and pulses all increase as people become wealthier in urban Bangladesh (Figure 4.10). Note that the income elasticity for pulses appears to be lower than for the other foods (i.e. the points in Figure 4.10 are closer together for pulses than for the other foods), consistent with the slow growth in pulse consumption at the subregional level noted earlier.

Figure 4.10 Consumption of various foods per month in urban Bangladesh by income group, 2010

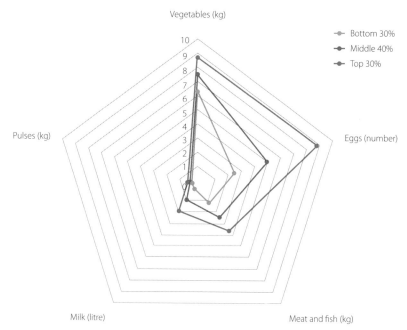

Source of raw data: Bangladesh Household Income and Expenditure survey (2010).

The importance of income as a key driver for food consumption holds true in both urban and rural areas and across countries, as can be seen by the data for meat and fish and milk consumption (Figure 4.11a and Figure 4.11b). In Bangladesh and Viet Nam, the income elasticity is more positive for meat and fish than it is for milk, while the reverse is true in India, reflecting different dietary preferences across countries (Natrajan and Jacob, 2018). Nevertheless, it is positive for all these commodities in all locations, and the same is true for eggs and vegetables (graphs not shown).

Figure 4.11a Consumption of meat and fish (kg) per month in various locations by income group

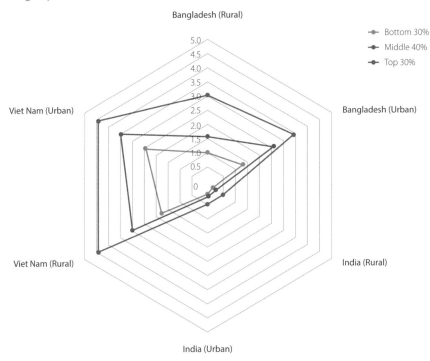

Figure 4.11b Consumption of milk (litres) per month in various locations by income group

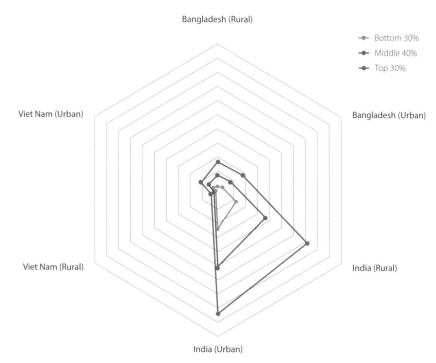

Source of raw data: Bangladesh Household Income and Expenditure Survey 2010, Bangladesh Bureau of Statistics, Ministry of Planning, Government of Bangladesh; India National Sample Survey 2009–2010, The National Sample Survey Organization, Government of India; Vietnam Household Living Standards Survey 2010, General Statistics Office, Government of Vietnam.

Although energy-dense staples (primarily rice) that lack key nutrients are still predominant in the diets of the poor, consumption of nutritious foods by the poor has been increasing over time, as can be seen in the case of egg consumption across three countries in both rural and urban areas (Figure 4.12). This is not surprising given that the incomes of the poor have been increasing (Chapter 2) and that consumption is positively correlated with income.

Figure 4.12 Consumption of eggs per month over time for the poorest 30 percent of the population

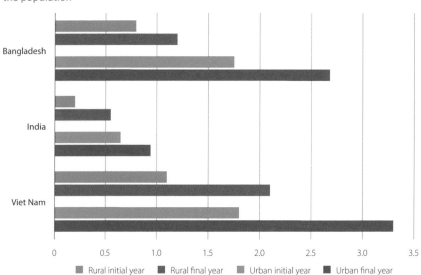

Sources of raw data: Bangladesh Household Income and Expenditure Survey 2005 and 2010, Bangladesh Bureau of Statistics, Ministry of Planning, Government of Bangladesh; India National Sample Survey 1993–1994 and 2009–2010, The National Sample Survey Organization, Government of India; Vietnam Household Living Standards Survey 2002 and 2010, General Statistics Office, Government of Vietnam.

The increased consumption of nutritious foods with the rising wealth conditions described above helps to explain both cross-country analyses of stunting and analyses using household survey data that show an inverse relationship between household wealth and the prevalence of stunting. Analyses of cross-country data have found that higher levels of per capita income are strongly associated with lower prevalence of stunting (Frongillo Jr., Onis and Hanson, 1997). Ruel and Alderman (2013) also found a negative relationship between the prevalence of stunting and the level of per capita GDP, but noted that that a given increase in GDP per capita was associated with a smaller reduction in stunting prevalence than in poverty.

Within countries, stunting rates are much higher for the poorest 20 percent than for the richest 20 percent (Figure 4.13). Finally, studies using household survey data that control for the effects of multiple independent variables have typically found measures of household wealth to be an important determinant of the prevalence of stunting (Cunningham *et al.*, 2017; Headey *et al.*, 2015; Headey and Hoddinott, 2016; Headey, Hoddinott and Park, 2017; Nisbett *et al.*, 2017b, 2017a; O'Donnell, Nicolás and Van Doorslaer, 2007; Raju and D'Souza, 2017; Zanello, Srinivasan and Shankar, 2016).

Figure 4.13 Prevalence of stunting by wealth quintile, average of 10 Asian developing countries

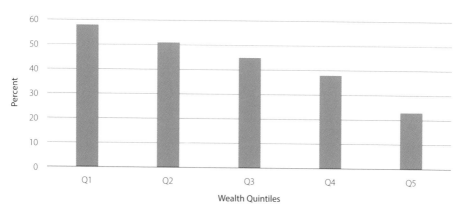

Source of raw data: Various Demographic and Health Surveys. Numbers are population-weighted. Countries included are: Bangladesh, Bhutan, Cambodia, India, Lao PDR, Mongolia, Nepal, Pakistan, Timor-Leste and Viet Nam. No data by quintile are available for Pacific countries.

While the impact of income on reducing undernutrition is obvious from Figure 4.13, it is also striking that stunting rates are so high in the top income quintile. This observation suggests that income is not the only determinant of undernutrition and that other factors are at work as well. Lack of nutrition education (e.g. knowledge surrounding which foods are nutritious, the importance of breastfeeding and the importance of early childhood nutrition for adult life) can lead to poor nutrition outcomes even when income is a minor constraint, as can lack of improved sanitation and clean drinking water.

Overall, the increased intake of nutritious foods in the region over time, along with better education, investments in clean drinking water and improved sanitation (Bhutta *et al.*, 2013), have led to substantial progress in reducing undernutrition over time, e.g. the prevalence of stunting in children under the age of five (Figure 4.2).

Thus, economic growth (provided it reaches the poor) and higher household incomes give households the financial resources (economic access) to consume more diverse diets, which are more expensive per calorie than staple foods and are not affordable to a substantial percentage of households in a range of countries in the region (WFP, 2015, 2017a, 2017b, 2017c, 2017d). Because animal-source foods contain high-quality protein, an appropriate level of consumption of these foods helps to reduce stunting (Headey, Hirvonen and Hoddinott, 2017). Furthermore, because of the importance of economic access in being able to afford more nutritious food, a more equal distribution of growth and wealth (holding constant the growth rate) will lead to more rapid improvements in nutritional outcomes. When economic and agricultural growth are slow, however, it is more difficult to reduce malnutrition (see Box 2).

Box 2 The impact of slow GDP and agricultural sector growth on diet in the Pacific islands

While obesity rates have risen worldwide over the past three decades, the greatest and most significant increase has occurred in the Pacific Island Countries (PICs), which contribute all of the world's ten most obese and overweight nations (NCD-RisC, 2016). Obesity rates in many of these countries surpass 60 percent, substantially higher than in other subregions. The resultant disability and early death have become an important public health, national economic, and regional political issue among the PICs (Pacific Islands Forum Secretariat, 2011). The obesity crisis is caused by many factors, but the nutrition transition towards refined foods high in sugar, salt and fats plays the major role in the current epidemic (Snowdon *et al.,* 2013). Low and stagnant consumption of fruits and vegetables may also play a role (Figure 4.7b).

The slow rate of growth in per capita incomes (Figure 2.1), coupled with especially slow growth in agriculture (Figure 2.14), has contributed to the rise in unhealthy diets. Slow growth in per capita income encourages people to purchase food that is cheaper but less nutritious. At the same time, the smallholder agriculture sector has limited capacity to supply the domestic market with locally produced fruit and vegetables at affordable prices because of low levels of electrification (see Chapter 5) and limited access to improved inputs. This has led to an increase in the relative price of domestic foods that have good nutritional value (Figure 4.14; ADB (2011); Evans *et al.* (2001)), encouraging a shift in the diet towards energy dense foods that are detrimental to long-term health if consumed in excessive amounts. In order to achieve sustainable improvements in nutritional status, it will be helpful to address the supply-side dimensions (see Chapter 6) affecting food purchasing decisions in the Pacific by creating a dynamic agricultural sector capable of producing nutritious foods at competitive and affordable prices. Given the relatively high share of calories coming from imports in the Pacific (Map 2.3b), nutrition-sensitive trade policies may also need to play a role.

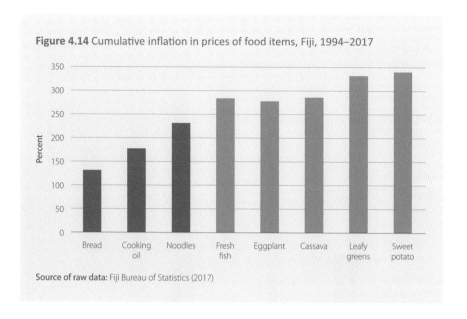

Figure 4.14 Cumulative inflation in prices of food items, Fiji, 1994–2017

Source of raw data: Fiji Bureau of Statistics (2017)

Food prices

Greater incomes are the most important determinant of economic access to nutritious food, but the price of food is also important. Since the turn of the twenty-first century, the food price subindex has increased faster than the overall consumer price index (CPI) in a wide range of countries in the region, indicating that food prices have risen relatively faster than the prices of other components of the CPI (the blue bars in Figure 4.15 are all positive, with only one exception). One likely reason for this increase in real food prices is that prices for a range of different food commodities on world markets hit all-time historic lows at the turn of the century, and have since increased. While price transmission from world to domestic markets is not immediate or automatic, especially in the Asia-Pacific region (Dawe, 2009), world prices do have eventual effects on domestic prices.

Within the food category, the prices of fruits and vegetables have risen more rapidly than the overall price of food in nearly all countries (the orange bars in Figure 4.15 are all positive except for Malaysia). The consistent increases in the relative prices of fruits and vegetables are especially remarkable because they have occurred over a sustained period of time (10 to 15 years). The sustained increases are consistent with consumers wanting to diversify their diets as their incomes grow, with their demand for fruits and vegetables increasing more rapidly than their demand for cereals and staple foods. At the same time, cultivation of fruits and vegetables is relatively labour-intensive (see Chapter 6), implying that cultivation will become more expensive as rural wages increase. These two forces point in the same direction: higher prices.

Figure 4.15 Relative price changes for food and fruits and vegetables, 2000–2016

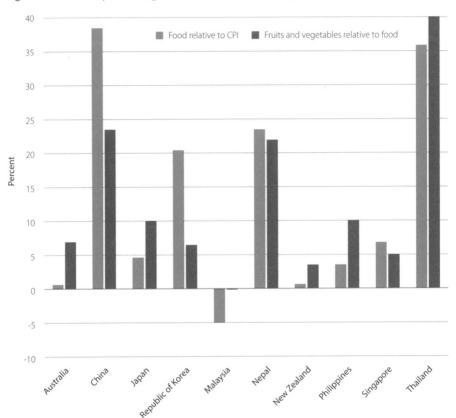

Note: The countries shown are all countries in the region for which eight or more years of recent data (ending in 2015 or 2016) were available on both the food component and the fruits and vegetables component of the consumer price index. The starting year is 2000 or the earliest year available. Beginning and ending years for analysis are as follows: Australia (2001, 2016), China (2005, 2015), Japan (2001, 2016), Republic of Korea (2000, 2016), Malaysia (2007, 2015), Nepal (2006/7, 2015/6), New Zealand (2001, 2016), Philippines (2000, 2016), Singapore (2001, 2016), Thailand (2001, 2016). Value for Thailand for fruits and vegetables relative to food is 148, but vertical axis is truncated in order to show other values more clearly.

The higher real prices for fruits and vegetables over time would be expected to reduce consumption of those items, all other things being equal. All other things are not equal, however, as incomes have risen rapidly in the region and consumption of fruits and vegetables has increased (Figure 4.7b). Thus, the higher prices for fruits and vegetables have not reduced actual consumption, but have caused consumption to be less than it otherwise would have been. For example, using a price elasticity of demand for fruits and vegetables in low-income countries of -1 (Regmi *et al.*, 2001), a 10 percent increase in the prices of these items would lead to a 10 percent decline in consumption (all other things being equal). The increased prices for fruits and vegetables make it more difficult for consumers to afford a nutritious diet, particularly the poor, because price elasticities are greater for the poor than for the rich (who are less responsive to prices). The high prices may particularly affect the urban poor, as the scarcity of land and the poor quality of water make it more difficult to have a home garden than in rural areas.

The price of the staple food also has an important impact on nutrition, especially for the poor, because it accounts for a substantial fraction of their total expenditures (in economics jargon, it has a substantial 'income' effect). When the price of the staple food is high, it can crowd out expenditures on other more nutritious foods such as eggs, milk and vegetables, simply because poor people have limited budgets and spend a higher share of their income on the staple food. Block *et al.* (2004) showed that when rice prices increased in Indonesia in the late 1990s, mothers in poor families responded by reducing their intake of dietary energy in order to better feed their children, leading to an increase in maternal wasting. Furthermore, purchases of more nutritious foods declined in order to afford the more expensive rice. This led to a measurable decline in blood haemoglobin levels in young children (and in their mothers), thus increasing the probability of developmental damage. A negative correlation between rice prices and nutritional status has also been observed in Bangladesh (Torlesse, Kiess and Bloem, 2003).

Urbanization

Urbanization can also contribute to increased consumption of certain foods, although not many studies try to isolate the impact of urbanization per se from the fact that urban residents tend to have greater incomes (Stage, Stage and Mcgranahan, 2010). In the case of Indonesia, Warr, Widodo and Yusuf (2018) found that urbanization, after controlling for income, leads to increased demand for chicken and eggs, but decreased demand for staples (rice and wheat). Their findings were similar to those of Huang and Bouis (2001) for Taiwan Province of the People's Republic of China, who found decreased overall demand for staples and increased demand for meat, fish and fruit. For vegetable oils, the urban-rural differentials may be smaller – Gaskell (2015) found a differential of only 5 percent in Indonesia after controlling for income.

There are likely several reasons for the impact of urbanization (after controlling for income) on the types of food eaten. Greater availability of refrigeration in urban areas, both at household level (Figure 4.16) and in value chains, may be part of the reason, as many of the foods whose consumption tends to increase are perishable. Several studies have found a correlation between meat consumption and refrigerator ownership after controlling for income (Gale et al., 2005; Lyon and Durham, 1999; OECD and FAO, 2013; Zhao and Thompson, 2013). Another possibility is that in urban areas, people come into closer contact with wealthier people, and try to emulate their consumption patterns.

However, growing incomes and urbanization can also lead to increased consumption of not just nutritious foods but also foods high in salt, sugar and saturated fat, especially in urban environments. Given traffic congestion, longer work hours, more women working outside the home and more formal work structures, people living in urban environments typically place a high value on the convenience of food preparation, leading to increased expenditures on processed foods (Warr, Widodo and Yusuf, 2018).

Globalization, especially the social dimensions (personal contacts, information flows), also appears to play a role in driving increased intake of dietary energy and a greater prevalence of obesity (Costa-Font and Mas, 2016), perhaps by influencing consumer preferences. These processes likely affect urban populations more than rural populations, given that urban areas are more exposed to globalization. Advertising might also affect consumer preferences in a similar way, and there is evidence that it may be more effective in urban areas than in rural areas, given the younger age structure of the population in the former (FAO, 2004).

Figure 4.16 Household ownership of refrigerators, urban and rural

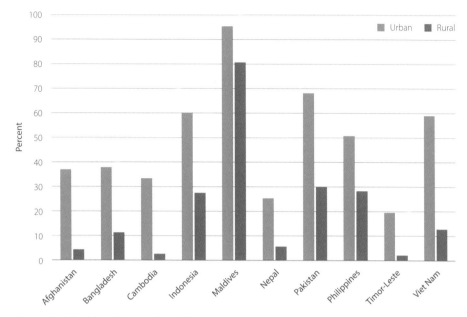

Source of raw data: Various Demographic Health Surveys (DHS)

Diversification of production to meet consumer demand

In line with the changes in diets towards animal-source foods and fruits and vegetables, the structure of production patterns at the national level is changing. In the early 1990s, rice ranked first in East, Southeast and South Asia in terms of farm production value (data for that time in the Pacific are sparse). But in recent years, rice has been replaced by pork in East Asia and by milk in South Asia, reflecting the shifts in diets towards more animal-source foods.

In most of the subregions, the share of cereals, roots and tubers in production value has declined over time (Figure 4.17) due to more rapid growth in livestock, aquaculture and fruits and vegetables. Furthermore, in all of the subregions, the combined shares of livestock, aquaculture, fruits and vegetables exceed that for cereals, roots and tubers. Despite the importance of the latter crops, it is clear that the agriculture sector produces much more than cereals, roots and tubers.

Within the category of cereals, rice production is still dominant, except in the Pacific, where root crops predominate. However, the share of maize in total cereal production value has generally increased (especially in East and South Asia), because maize supplies the energy for most livestock feed – thus,

Figure 4.17 Shares of various products in total farm production value, 1990s and 2016

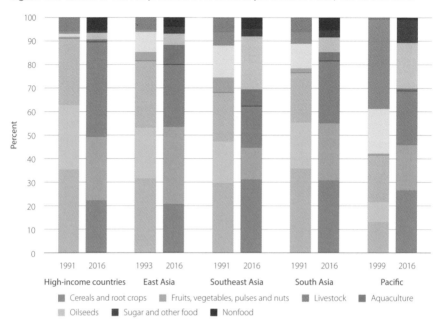

Sources of raw data: FAO (2018b), FAO (2018c)

Note: Starting year for each subregion varies depending upon data availability.

input demand for maize increases along with increased consumer demand for livestock products. Minor cereals remain just that – minor. There has been some small increase in the share of minor cereals in total cereal production value in the high-income countries, nearly all of which has been due to increased barley production in Australia, as well as, to a much lesser extent, increased sorghum production for animal feed. But in the other subregions, the share of minor cereals in total cereals production value has declined during the past 20 years.

Just as the share of cereals, roots and tubers in total production value has generally declined, the shares of fruits and vegetables, livestock and aquaculture have all generally increased. Southeast Asia is an exception to some extent, as the rapid growth of oil palm due to a strong comparative advantage and a continuing emphasis on rice production in some countries has overwhelmed the increases in livestock and fruits and vegetables. The shift towards livestock and fruits and vegetables is also evident in changes in area harvested, with China providing the most striking example. In China, the area harvested[1] to

---

1   'Area harvested' incorporates cropping intensity (the number of crops grown per year) and physical area. Thus, if a farmer grows two crops of rice in one year on one hectare, that counts as two hectares of rice area harvested.

fruits and vegetables now exceeds that for any single cereal (Figure 4.18). The harvested maize area, used primarily for animal feed, now exceeds the areas for rice and wheat.

Crop diversification at the national level does not necessarily imply crop diversification at the level of individual farms. In fact, national diversification is consistent with increasing specialization at the farm level, with different farms specializing in different crops. In Thailand, farmers have been increasingly specializing in a smaller number of crops during the past 10 to 15 years, a trend that manifests itself across all the different income classes. Farmers that raise livestock and fish have also become more specialized in particular animal species (Poapongsakorn, Pantakua and Wiwatvicha, 2016).

While there can be ecological benefits to more diverse landscapes, it is difficult to optimize the spatial scale at which such diversification needs to occur (e.g. farm, village, watershed), especially because there are increasing incentives for individual farmers to specialize in particular crops as farming becomes more knowledge-intensive. The increasing role of non-farm incomes in rural areas also encourages farm specialization. By providing an alternative source of income that is less dependent on the vagaries of the weather, it helps farm households to manage agricultural risk, thus reducing the need for farms to plant different crops in order to manage that risk.

Figure 4.18 Area harvested to different crops in China, 1962–2016

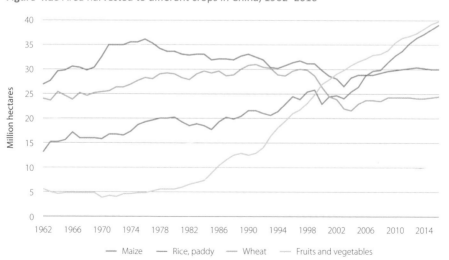

Source of raw data: FAO (2018b)

Producing more high-value foods such as livestock, fish and fruits and vegetables will be important for improving both farm incomes and nutritional outcomes. As noted above, these production shifts are taking place, but not rapidly enough. For example, prices of fruits and vegetables are increasing over time in a range of countries (Figure 4.15), showing that demand is growing faster than supply and making it more difficult for the poor to afford them.

Farmers may be finding it difficult to grow enough fruits and vegetables for a number of reasons (Kasem and Thapa, 2011). First, growing new crops requires new knowledge that farmers may not have. Second, not all land is suitable for growing fruits and vegetables – agricultural production choices are heavily conditioned by local climate, water supply, topography and soil, most of which cannot be changed. Third, many fruits require several years before the tree produces a harvest, and farmers need an alternative source of income during this transition period. Fourth, growing fruits and vegetables tends to be riskier than for rice, with greater fluctuations in both production and prices. Fifth, fruit and vegetable production is more labour-intensive than rice (Alviola, Cataquiz and Francisco, 2002; Kasem and Thapa, 2011; Maertens, Minten and Swinnen, 2012). This means growing production costs in countries where wages are rising, although higher output prices should compensate for this (Zhong, 2014). The increased labour intensity also means more employment for the rural poor (Dawe, 2006). Sixth, rice import restrictions in some countries raise domestic rice prices substantially, thus discouraging farmers from diversifying into other crops. Overcoming these numerous barriers will require innovations in institutions and policies.

Organic production is another high-value option for farmers that has been growing rapidly in many countries, especially in the Pacific and in high-income countries, where it often has a share of between 1 and 5 percent of the total agricultural area. However, in the developing countries of East, Southeast and South Asia, it is still very much a niche market, accounting for less than 1 percent of the total agricultural areas in nearly all countries in those subregions. The low market share in these subregions is most likely due to the higher prices for organic food.

Summary

There has been considerable progress in reducing the incidence of undernutrition due to economic growth, better education and improved water and sanitation infrastructure, but it remains a substantial problem in much of the region. Economic growth and urbanization have led to increased consumption of animal-source foods, fruits and vegetables, including for the poor, but have also led to increased intake of fats, sugar and salt as part of the nutrition transition. Coupled with declining levels of physical activity due to urbanization, these trends are leading to rising obesity. Per capita consumption of staple foods is declining in much of the region. Farming systems are shifting to accommodate all of these changing demands, with the livestock and aquaculture sectors increasingly accounting for a larger share of the value of agricultural output. However, production systems are not changing rapidly enough to prevent price increases for some healthy foods, particularly fruits and vegetables.

References

ADB. 2011. *Food security and climate change in the Pacific: Rethinking the options.* Mandaluyong City, ADB. (also available at http://hdl.handle.net/11540/953).

Allen, L.H. 2008. To what extent can food-based approaches improve micronutrient status? *Asia Pacific Journal of Clinical Nutrition*, 17(S1): 103–105. https://doi.org/10.12691/ajfn-5-1-1

Allotey, P., Davey, T. & Reidpath, D.D. 2014. NCDs in low and middle-income countries – assessing the capacity of health systems to respond to population needs. *BMC Public Health*, 14(Suppl 2). https://doi.org/10.1186/1471-2458-14-S2-S1

Alviola, P.A., Cataquiz, G.C. & Francisco, S. 2002. Global competitiveness of rice-vegetable farming systems: Implication to Philippine food security. *Paper presented at the International Rice Research Conference, 16–20 September 2002, Beijing, China.*

Baker, P. & Friel, S. 2016. Food systems transformations, ultra-processed food markets and the nutrition transition in Asia. *Globalization and Health*, 12(80). https://doi.org/10.1186/s12992-016-0223-3

Bernadette, B., Munthe, C. & Taylor, M. 2016. RPT-cooked in Indonesia: Phoney rice data threatens food supply, 25 January 2016. (also available at https://www.reuters.com/article/indonesia-rice-data/rpt-cooked-in-indonesia-phoney-rice-data-threatens-food-supply-idUSL3N15701Q).

Bhurosy, T. & Jeewon, R. 2014. Overweight and obesity epidemic in developing countries: a problem with diet, physical activity, or socioeconomic status? *Scientific World Journal* (964236). https://doi.org/10.1155/2014/964236

Bhutta, Z.A., Das, J.K., Rizvi, A., Gaffey, M.F., Walker, N., Horton, S., Webb, P., Lartey, A. & Black, R.E. 2013. Evidence-based interventions for improvement of maternal and child nutrition: what can be done and at what cost? *The Lancet*, 382: 452–477. https://doi.org/10.1016/S0140-6736(13)60996-4

Black, R.E., Victora, C.G., Walker, S.P., Bhutta, Z.A., Christian, P., De Onis, M., Ezzati, M., *et al.* 2013. Maternal and child undernutrition and overweight in low-income and middle-income countries. *The Lancet*, 382: 427–451. https://doi.org/10.1016/S0140-6736(13)60937-X

Block, S.A., Kiess, L., Webb, P., Kosen, S., Moench-Pfanner, R., Bloem, M. & Timmer, P. 2004. Macro shocks and micro outcomes: child nutrition during Indonesia's crisis. *Economics and Human Biology*, 2: 21–44. https://doi.org/10.1016/j.ehb.2003.12.007

Costa-Font, J. & Mas, N. 2016. 'Globesity'? The effects of globalization on obesity and caloric intake. *Food Policy*, 64: 121–132. https://doi.org/10.1016/j.foodpol.2016.10.001

Cunningham, K., Headey, D., Singh, A., Karmacharya, C. & Rana, P.P. 2017. Maternal and child nutrition in Nepal: Examining drivers of progress from the mid-1990s to 2010s. *Global Food Security*, 13: 30–37. https://doi.org/10.1016/j.gfs.2017.02.001

Dawe, D. 2006. Rice trade liberalization will benefit the poor. In D. Dawe, P.F. Moya & C.B. Casiwan, eds. *Why does the Philippines import rice? Meeting the challenge of trade liberalization*, pp. 43–52. IRRI & PhilRice. (also available at http://irri.org/resources/publications/books/item/why-does-the-philippines-import-rice).

Dawe, D. 2009. Cereal price transmission in several large Asian countries during the global food crisis. *Asian Journal of Agriculture and Development*, 6(1): 1–12. (also available at http://ageconsearch.umn.edu/record/199068?ln=en).

Evans, M., Sinclair, R.C., Fusimalohi, C. & Liava'a, V. 2001. Globalization, diet, and health: an example from Tonga. *Bulletin of the World Health Organization*, 79(9): 856–862. https://doi.org/10.1590/S0042-96862001000900011

FAO. 2004. Globalization of food systems in developing countries: impact on food security and nutrition. *FAO Food and Nutrition Paper*, 83. (also available at http://www.fao.org/docrep/007/y5736e/y5736e00.htm).

FAO. 2013. *The state of food and agriculture: Food systems for better nutrition*. Rome. (also available at http://www.fao.org/publications/sofa/2013/en/).

FAO. 2016. *Regional overview of food insecurity: Asia and the Pacific*. Bangkok, Regional Office for Asia and the Pacific. (also available at http://www.fao.org/publications/card/en/c/ea009918-8124-4320-a5d9-0a296b07c141/).

FAO. 2018a. *Asia and the Pacific regional overview of food security and nutrition*. Bangkok, FAO Regional Office for Asia and the Pacific. (also available at http://www.fao.org/3/CA0950EN/CA0950EN.pdf).

FAO. 2018b. *FAOSTAT* [online]. www.fao.org/faostat/

FAO. 2018c. *AQUASTAT* [online]. http://www.fao.org/nr/water/aquastat/main/index.stm

Fiji Bureau of Statistics. 2017. *Consumer price index records* [online]. https://www.statsfiji.gov.fj/

Frongillo Jr., E.A., Onis, M. de & Hanson, K.M.P. 1997. Socioeconomic and demographic factors are associated with worldwide patterns of stunting and wasting of children. *The Journal of Nutrition*, 127(12): 2302–2309. https://doi.org/10.1093/jn/127.12.2302

Gale, F., Tang, P., Bai, X. & Xu, H. 2005. Commercialization of food consumption in rural China. *Economic Research Report*, (also available at http://handle.nal.usda.gov/10113/18039).

Gaskell, J.C. 2015. The role of markets, technology, and policy in generating palm-oil demand in Indonesia. *Bulletin of Indonesian Economic Studies*, 51: 29–45. https://doi.org/10.1080/00074918.2015.1016566

Hawkes, C., Smith, T.G., Jewell, J., Wardle, J., Hammond, R.A., Friel, S., Thow, A.M. & Kain, J. 2015. Smart food policies for obesity prevention. *The Lancet*, 385: 2410–2421. https://doi. org/10.1016/S0140-6736(14)61745-1

Headey, D., Hirvonen, K. & Hoddinott, J. 2017. Animal sourced foods and child stunting. *IFPRI Discussion Paper 01695*, Washington, DC. (also available at http://ebrary.ifpri.org/ cdm/ref/collection/p15738coll2/id/132232).

Headey, D., Hoddinott, J., Ali, D., Tesfaye, R. & Dereje, M. 2015. The other Asian enigma: Explaining the rapid reduction of undernutrition in Bangladesh. *World Development*, 66: 749–761. https://doi.org/10.106/j.worlddev.2014.09.022

Headey, D., Hoddinott, J. & Park, S. 2017. Accounting for nutritional changes in six success stories: A regression-decomposition approach. *Global Food Security*, 13: 12–20. https://doi.org/10.1016/j.gfs.2017.02.003

Headey, D.D. & Hoddinott, J. 2016. Agriculture, nutrition and the green revolution in Bangladesh. *Agricultural Systems*, 149: 122–131. https://doi.org/10.1016/j.agsy.2016.09.001

Helble, M. & Francisco, K. 2017. The imminent obesity crisis in Asia and the Pacific: First cost estimates. ADBI Working Paper Series No. 743. Tokyo. (also available at https://www. adb.org/publications/imminent-obesity-crisis-asia-and-pacific-first-cost-estimates).

Hess, S.Y. 2017. Zinc deficiency. In S. de Pee, D. Taren & M.W. Bloem, eds. *Nutrition and health in a developing world.* Third edition, pp. 265–285. New York, Humana Press.

Hoddinott, J., Behrman, J.R., Maluccio, J.A., Melgar, P., Quisumbing, A.R., Ramirez-zea, M., Stein, A.D., et al. 2013. Adult consequences of growth failure in early childhood. *American Journal of Clinical Nutrition*, 98(5): 1170–1178. https://doi.org/10.3945/ ajcn.113.064584

Hoddinott, J.F. 2013. The economic cost of malnutrition. In M. Eggersdorfer, K. Kraemer, M. Ruel, M. Van Ameringen, H.K. Biesalski, M. Bloem, J. Chen, A. Lateef & V. Mannar, eds. *The road to good nutrition: a global perspective*, pp. 64–73. Basel. (also available at http://www.ifpri.org/blog/road-good-nutrition).

Huang, J. & Bouis, H. 2001. Structural changes in the demand for food in Asia: Empirical evidence from Taiwan. *Agricultural Economics*, 26(1): 57–69. https://doi.org/10.1111/j.1574-0862.2001.tb00054.x

IFPRI. 2015. Global nutrition report 2015: Actions and accountability to advance nutrition and sustainable development. Washington, DC. (also available at http://www.ifpri.org/ publication/global-nutrition-report-2015).

IFPRI. 2016. Global nutrition report 2016: From promise to impact – Ending malnutrition by 2030. Washington, DC. (also available at http://www.ifpri.org/publication/global-nutrition-report-2016-promise-impact-ending-malnutrition-2030)

Kasem, S. & Thapa, G.B. 2011. Crop diversification in Thailand: Status, determinants, and effects on income and use of inputs. *Land Use Policy*, 28: 618–628. https://doi.org/10.1016/j. landusepol.2010.12.001

Lyon, C. & Durham, C. 1999. Refrigeration and food demand in China: Can refrigerator ownership help predict consumption of food products in China? *Paper presented at the Chinese Agriculture and WTO, Proceedings of the WCC-101 (December 2–3, 1999).*

Ma, R.C.W. & Chan, J.C.N. 2013. Type 2 diabetes in East Asians: Similarities and differences with populations in Europe and the United States. *Annals of the New York Academy of Sciences*, 1281: 64–91. https://doi.org/10.1111/nyas.12098

Maertens, M., Minten, B. & Swinnen, J. 2012. Modern food supply chains and development: Evidence from horticulture export Sectors in Sub-Saharan Africa. *Development Policy Review*, 30(4): 473–497. https://doi.org/10.1111/j.1467-7679.2012.00585.x

Micronutrient Initiative. 2015. *Micronutrient Initiative* [online]. www.micronutrient.org

Monteiro, C., Conde, W., Lu, B. & Popkin, B. 2004. Obesity and inequities in health in the developing world. *International Journal of Obesity*, 28: 1181–1186. https://doi.org/10.1038/sj.ijo.0802716

Natrajan, B. & Jacob, S. 2018. 'Provincialising' vegetarianism: Putting Indian food habits in their place. *Economic & Political Weekly*, LIII(9): 54–64. (also available at https://www.epw.in/journal/2018/9).

NCD Risk Factor Collaboration (NCD-RisC). 2016. Trends in adult body-mass index in 200 countries from 1975 to 2014: a pooled analysis of 1698 population-based measurement studies with 19.2 million participants. *The Lancet*, 387: 1377–1396. https://doi.org/10.1016/S0140-6736(16)30054-X

NCD Risk Factor Collaboration (NCD-RisC). 2018. *NCD-RisC Data and Publications* [online]. http://ncdrisc.org/

Nisbett, N., Bold, M. van den, Menon, S.G., Davis, P., Roopnaraine, T., Kampman, H., Kohli, et al. 2017a. Community-level perceptions of drivers of change in nutrition: Evidence from South Asia and sub-Saharan Africa. *Global Food Security*, 13: 74–82. https://doi.org/10.1016/j.gfs.2017.01.006

Nisbett, N., Davis, P., Yosef, S. & Akhtar, N. 2017b. Bangladesh's story of change in nutrition: Strong improvements in basic and underlying determinants with an unfinished agenda for direct community level support. *Global Food Security*, 13: 21–29. https://doi.org/10.1016/j.gfs.2017.01.005

O'Donnell, O., Nicolás, Á.L. & Van Doorslaer, E. 2007. Growing richer and taller: Explaining change in the distribution of child nutritional status during Vietnam's economic boom. *Tinbergen Institute Discussion Paper TI 2007–2008/3*, Amsterdam.

OECD & FAO. 2013. *OECD-FAO Agricultural Outlook 2013–2022*. OECD Publishing. (also available at https://www.oecd-ilibrary.org/agriculture-and-food/oecd-fao-agricultural-outlook-2013_agr_outlook-2013-en).

Pacific Islands Forum Secretariat. 2011. Joint Statement of the Pacific Island Forum Leaders and United Nations Secretary General. *42nd Pacific Island Forum, 7–8 September 2011, Auckland, New Zealand.* (also available at http://sdg.iisd.org/news/42nd-pacific-island-forum-develops-communique-and-joint-statement-from-leaders-and-un-secretary-general/).

de Pee, S. 2017. Nutrient needs and approaches to meeting them. In S. de Pee, D. Taren & M.W. Bloem, eds. *Nutrition and health in a developing world*. Third edition, pp. 159–180. New York, Humana Press.

Poapongsakorn, N., Pantakua, K. & Wiwatvicha, S. 2016. The structural and rural transformation in selected Asian countries

Popkin, B. 2006. *What is the nutrition transition?* [online]. http://www.cpc.unc.edu/projects/nutrans/whatis

Raju, D. & D'Souza, R. 2017. Child undernutrition in Pakistan: What do we know? Policy Research Working Paper No. 8049. Washington, DC, World Bank Group. (also available at http://documents.worldbank.org/curated/en/810811493910657388/Child-undernutrition-in-Pakistan-what-do-we-know).

Regmi, A., Deepak, M.S., Seale Jr., J.L. & Bernstein, J. 2001. Cross-country analysis of food consumption patterns. *Changing Structure of Global Food Consumption and Trade*, WRS-01-1: 14–22. (also available at https://www.ers.usda.gov/publications/pub-details/?pubid=40319).

Ruel, M.T. & Alderman, H. 2013. Nutrition-sensitive interventions and programmes: how they help to accelerate progress in improving maternal and child nutrition? *The Lancet*, 382(9891): 536–551. https://doi.org/http://dx.doi.org/10.1016/S0140-6376(13)60843-0

SKRI. 2016. *President Jokowi: Only BPS is responsible* for data [online]. http://setkab.go.id/en/president-jokowi-only-bps-is-responsible-for-data/

Snowdon, W., Raj, A., Reeve, E., Guerrero, R.L., Fesaitu, J., Cateine, K. & Guignet, C. 2013. Processed foods available in the Pacific Islands. *Globalization and Health*, 9(53). https://doi.org/10.1186/1744-8603-9-53

Stage, J., Stage, J. & Mcgranahan, G. 2010. Is urbanization contributing to higher food prices? *Environment & Urbanization*, 22(1): 199–215. https://doi.org/10.1177/0956247809359644

Stevens, G.A., Bennett, J.E., Hennocq, Q., Lu, Y., De-Regil, L.M., Rogers, L., Danaei, G., Li, G., *et al.* 2015. Trends and mortality effects of vitamin A deficiency in children in 138 low-income and middle-income countries between 1991 and 2013: a pooled analysis of population-based surveys. *Lancet Global Health*, 3: e528–536. https://doi.org/10.1016/S2214-109X(15)00039-X

Timmer, P. 2018. State-level structural transformation and poverty reduction in Malaysia: a multi-commodity approach

Timmer, P., Block, S.A. & Dawe, D. 2010. Long-run dynamics of rice consumption, 1960–2050. In S. Pandey, D. Byerlee, D. Dawe, A. Dobermann, S. Mohanty, S. Rozelle & B. Hardy, eds. *Rice in the Global Economy: Strategic Research and Policy Issues for Food Security*, Los Banos, International Rice Research Institute. (also available at http://irri.org/resources/publications/books/rice-in-the-global-economy-strategic-research-and-policy-issues-for-food-security).

Torlesse, H., Kiess, L. & Bloem, M.W. 2003. Association of household rice expenditure with child nutritional status indicates a role for macroeconomic food policy in combating malnutrition. *The Journal of Nutrition*, 133(5): 1320–1325. https://doi.org/10.1093/jn/133.5.1320

UNICEF, WHO & The World Bank Group. 2018. *Joint child malnutrition estimates* [online]. http://datatopics.worldbank.org/child-malnutrition/

Victora, C.G., Adair, L., Fall, C., Hallal, P.C., Martorell, R., Richter, L. & Sachdev, H.S. 2008. Maternal and child undernutrition: Consequences for adult health and human capital. *The Lancet*, 371: 340–357. https://doi.org/10.1016/S0140-6736(07)61692-4

Warr, P., Widodo, M.A. & Yusuf, A.A. 2018. Urbanization and the Demand for Food in Indonesia

Wen, C.P., Cheng, T.Y.D., Tsai, S.P., Chan, H.T., Hsu, H.L., Hsu, C.C. & Eriksen, M.P. 2009. Are Asians at greater mortality risk for being overweight than Caucasians? Redefining obesity for Asians. *Public Health Nutrition*, 12(4): 497–506. https://doi.org/10.1017/S1368980008002802

World Health Organization (WHO). 2018. Increasing fruit and vegetable consumption to reduce the risk of noncommunicable diseases. In: *e-Library of Evidence for Nutrition Actions (eLENA)* [online]. http://www.who.int/elena/titles/fruit_vegetables_ncds/en/

World Health Organization (WHO) expert consultation. 2004. Appropriate body-mass index for Asian populations and its implications for policy and intervention strategies. *The Lancet*, 363(1): 157–163. https://doi.org/10.1016/S0140-6736(03)15268-3

World Bank. 2011. *Bangladesh – Household income and expenditure survey: key findings and results 2010 (English)*. Washington DC.

World Food Programme (WFP). 2015. Sri Lanka – Minimum cost of nutritious diet. (October, 2015). (also available at https://www.wfp.org/content/sri-lanka-minimum-cost-nutritious-diet-october-2013-september-2014-october-2015).

World Food Programme (WFP). 2017a. Fill the nutrient gap Lao PDR. (also available at https://www.wfp.org/content/2017-fill-nutrient-gap?_ ga=2.86790555.2011256344.1518424351-1082568425.1480515367).

World Food Programme (WFP). 2017b. Fill the nutrient gap Pakistan. (also available at https://www.wfp.org/content/2017-fill-nutrient-gap?_ga=2.86790555.2011256344. 1518424351-1082568425.1480515367).

World Food Programme (WFP). 2017c. The cost of the diet study in Indonesia. (also available at https://www.wfp.org/content/indonesia-cost-diet-study).

World Food Programme (WFP). 2017d. Fill the nutrient gap Cambodia. (also available at https://www.wfp.org/content/2017-fill-nutrient-gap?_ga=2.86790555.2011256344. 1518424351-1082568425.1480515367).

World Health Organization (WHO). 2018. *Global health observatory data repository* [online]. http://apps.who.int/gho/data/view.main.v100230?lang=en

Young, M.F. & Ramakrishnan, U. 2017. Iron. In S. de Pee, D. Taren & M.W. Bloem, eds. *Nutrition and health in a developing world*, pp. 235–263. New York, Humana Press.

Zanello, G., Srinivasan, C.S. & Shankar, B. 2016. What explains Cambodia's success in reducing child stunting – 2000–2014? *PLoS ONE*, 11(9): e0162668. https://doi.org/10.1371/ journal.pone.0162668

Zhao, J. & Thompson, W. 2013. The effect of refrigerator use on meat consumption in rural China. *Paper presented at the Annual Meeting of Southern Agricultural Economics Association (February 2–5, 2013).* (also available at https://ideas.repec.org/s/ags/saea13.html).

Zhong, F. 2014. Impact of demographic change on agricultural mechanization: Farmers' adaptation and implication for public policy. *Paper presented at the NSD/IFPRI workshop on mechanization and agricultural transformation in Asia and Africa (June 18–19, 2014).* (also available at https://www.slideshare.net/IFPRIDSG/impact-of-demographic-change-on-agricultural-mechanization-farmers-adaptation-and-implication-for-public-policy).

5

TRANSFORMATION OF VALUE CHAINS

The increasing complexity of value chains

Value chains perform essential functions that ultimately allow consumers to eat the food produced by farmers: aggregation, storage, transport, processing and distribution. Structural transformation of the economy and associated changes in consumer demand have led to these value chain functions becoming increasingly complex. The increased sophistication of value chains has allowed for greater dietary diversity, while simultaneously introducing new challenges in ensuring that everyone can access healthy, sustainable diets. Governments require a good understanding of how food is transformed from farm to fork and why and how people are making their food choices in order to better tailor policies that enable the production and consumption of more nutritious food.

In performing these functions, value chains should be inclusive, efficient and nutrition-sensitive. Inclusive value chains facilitate the participation of small farm households, the rural landless and the urban poor, and both men and women so that they can improve their livelihoods and nutrition. For example, rural roads, cold chains and e-commerce on the Internet can encourage farmers to produce and sell greater quantities of production by opening up new markets. This same infrastructure can also provide lower prices for the urban (and rural) poor. Improved value chains allow farms to diversify and grow or raise higher value-added products (fruits, vegetables, dairy products, livestock, fish), which can create entrepreneurial opportunities that lead to new agribusiness operations and wage employment for farmers, other rural dwellers and urban residents.

Efficient value chains reduce aggregation, storage, transport and distribution costs incurred in moving produce from the farmgate to the market. These cost reductions allow higher prices for farmers and lower, more affordable prices for consumers, which are of particular importance for the poor. For example, improved road infrastructure can reduce transport times, cutting costs and lowering losses of perishable items. When value chains lower food losses and waste, they not only reduce marketing margins, but also help to improve the environment by ensuring that natural resources are not used to produce food that is never consumed.

Nutrition-sensitive value chains help to improve the delivery of nutritious food to consumers. As many nutritious foods are more perishable than staple grains, cold chains are necessary to make sure that the food reaches the consumer without having wilted or rotted. More efficient value chains can also be nutrition-sensitive by helping to avoid the deterioration of nutrient content, which declines in certain foods over time or when exposed to the elements.

Expansion of infrastructure

Inclusive, efficient and nutrition-sensitive value chains depend on a solid base of physical infrastructure. During the past 20 to 30 years, there has been a steady expansion of physical infrastructure throughout the region, although some subregions have more than others. This expansion has facilitated improved transportation, communication and sharing of knowledge.

Roads, especially farm-to-market roads in rural areas, are essential components of value chains. Urban transport infrastructure can also affect the efficiency of value chains – food marketing costs will be higher if there is substantial traffic congestion around cities. Road density has increased in nearly all countries, although comparable international data are sparse and not up to date.[1] For example, between 2000 and 2011 (the most recent data available), the number of kilometres of roads increased by more than 8 percent per annum in China and Malaysia (UNESCAP, 2018). Other countries have seen less rapid expansion, however.

Rural electrification is another crucial component of value chains, especially for perishable produce. The percentage of the rural population that has access to electricity has been steadily increasing in all subregions, and is now about 90 percent in the developing countries of East and Southeast Asia

---

1 SDG indicator 9.1.1 is the proportion of the rural population that lives within 2 kilometres of an all-season road. However, no data for this indicator are currently available and its methodology is under development.

(Figure 5.1). South Asia has been catching up in recent years (the percentage of the rural population with electricity doubled in 20 years), but the Pacific countries lag far behind, which makes it difficult for farmers to participate in the most remunerative value chains (especially when coupled with their geographic isolation). It also makes it more difficult for rural populations in those countries to have access to a wide variety of fresh nutritious food throughout the year. Urban electrification is at higher levels, approaching 100 percent in all subregions other than the Pacific, where it is above 80 percent.

Figure 5.1 Percentage of rural population with access to electricity by subregion, 1990–2016

Source of raw data: World Bank (2018)

A robust communications network is important to facilitate timely interactions between buyers and sellers of agricultural products. Before the era of mobile telephones, many households had only limited access to fixed-line telephones, with many people having to wait years for installation, even in urban areas. Since the turn of the century, however, the number of mobile telephone subscriptions per 100 people has expanded substantially in developing countries, and has reached more than 100 in Southeast Asia and the high-income countries (Figure 5.2).

Figure 5.2 Number of mobile telephone subscriptions per 100 people by subregion, 1990–2016

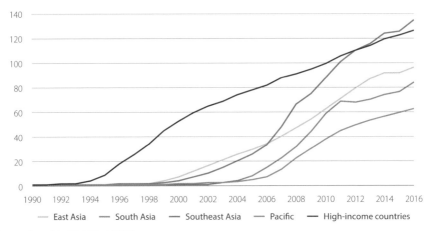

— East Asia — South Asia — Southeast Asia — Pacific — High-income countries

Source of raw data: World Bank (2018)

Figure 5.3 Percentage of population using the Internet by subregion, 2001–2016

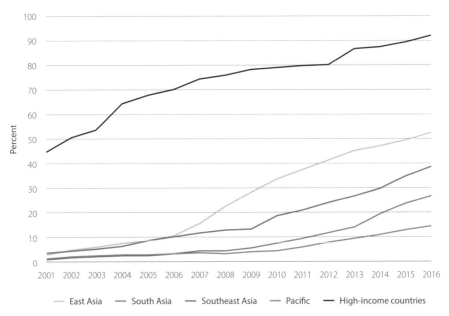

— East Asia — South Asia — Southeast Asia — Pacific — High-income countries

Source of raw data: World Bank (2018)

The Internet greatly facilitates access to new information in a manner that was unthinkable several decades ago. The percentage of the population using the Internet has increased substantially since 2000, although more than half the population in Southeast Asia, South Asia and the Pacific is still not using it (Figure 5.3). Here, there is still ample scope for further expansion that could facilitate the dissemination of knowledge surrounding good agricultural practices and help to connect consumers and producers directly.

Trends in value chains

In addition to the improved supply of infrastructure noted above, value chains are also being shaped by changing consumer demand due to greater incomes and urbanization. For example, rising wages and increased traffic congestion in urban areas are increasing demand for more convenient food preparation. On the domestic supply side, in response to consumer demand and the greater supply of public goods noted above, food manufacturing industries are becoming more important relative to primary agriculture, transport and storage networks that move food from farm to retail outlets are becoming more sophisticated and retail market outlets are growing increasingly diverse. International trade is also growing in importance.

Urbanization and demands for more convenient food preparation

People living in urban areas face more traffic congestion, live a longer distance from the workplace and have more rigid work schedules and longer working hours, relative to rural areas. Because of high land prices in urban areas, it is becoming more common for housing to devote less space to kitchens in an effort to conserve room for other functions. In tropical cities, the high cost of air-conditioning has encouraged people to eat out more often (Sahakian, M.; Saloma, C.; Erkman, 2017). These characteristics of urban areas are powerful driving forces leading people to demand more convenient food preparation.

The demand for more convenient food preparation manifests itself in several ways. One example is increased purchases of highly processed food in urban areas (Warr, Widodo, & Yusuf, 2018; Figure 5.4). Processed food is not bad in and of itself. It responds to clear consumer demands, but has advantages and disadvantages. Processed foods reduce food preparation time at home and allow for reduced food waste through the use of preservatives. As women are often responsible for food preparation, processed foods can also give women more flexibility to engage in formal employment. However, such foods are often high in added salt, sugar and saturated fats, which can have adverse health impacts (see Chapter 4).

Figure 5.4 Sources of food in urban and rural areas

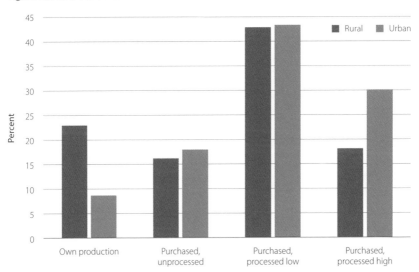

Source of raw data: Reardon *et al.* (2014)

Note: Countries included are Bangladesh, Indonesia, Nepal and Viet Nam.

The increased opportunity cost of time that leads to demand for more convenience can also encourage greater food waste. In the interests of convenience, people may purchase more food than is needed in order to avoid running out. Retail establishments may follow a similar logic in order to avoid running out of specific food items and alienating consumers.

Another way in which demand for more convenient food preparation expresses itself is as demand for eating meals prepared outside the home. These meals can be consumed at full service restaurants, fast food outlets or street stalls, or can be purchased in these outlets (or at supermarkets) and consumed at home. The share of food expenditures on food prepared outside the home has grown rapidly over the past 10 to 15 years in East and Southeast Asia. Eating away from home is much more common in urban areas than in rural areas in these regions (see Figure 5.5 for the case of Viet Nam, which is similar to the patterns for China, Indonesia, the Philippines and Thailand). Wealthier people tend to have a greater percentage of their expenditures on food prepared away from home, but this share is becoming substantial for the poor as well. For example, in Viet Nam, the poorest quintile more than doubled its share of total food expenditures on food prepared away from home between 2002 and 2014 (Figure 5.5). Eating food prepared away from home has been slower to take hold in South Asia, however, with such food accounting for a much lower share of total food expenditures (Figure 5.6; no data are available for any Pacific countries).

Figure 5.5 Percentage of total food expenditures spent on food prepared away from home in urban and rural areas of Viet Nam (2002 and 2014), by expenditure quintile

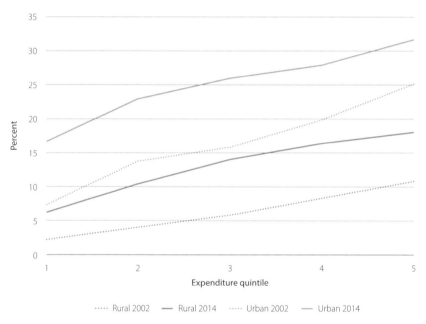

Source of raw data: Various Viet Nam Household Living Standard Surveys.

Figure 5.6 Percentage of total food expenditures spent on food prepared away from home in urban areas by expenditure quintiles

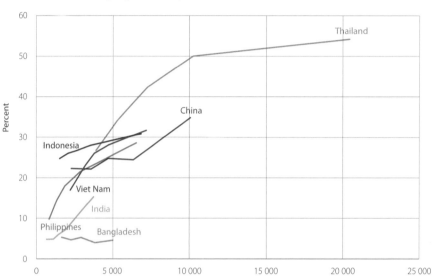

Source of raw data: Various household surveys: Bangladesh (2010), China (2009/10), India (2011/12), Indonesia (2014), Philippines (2012), Thailand (2015), Viet Nam (2014).

Note: Horizontal axis is in purchasing power parity (PPP) terms.

Consumers obtain food prepared away from home in a wide variety of ways. The relative frequency with which people use different types of outlets varies considerably from country to country (Figure 5.7). In China, fast food restaurants are quite common, while in Indonesia full-service restaurants have a larger market share. There is a heavier reliance on street food vendors in India, while in Malaysia there is more balance across the different types of outlets. For children, obtaining meals at school can be quite common in some countries (e.g. China; Euromonitor International, 2017) Emerging methods include a growing use of e-commerce, where consumers order food online for home delivery from restaurants and vendors.

Figure 5.7 Number of transactions per capita per year at various food outlets, national average

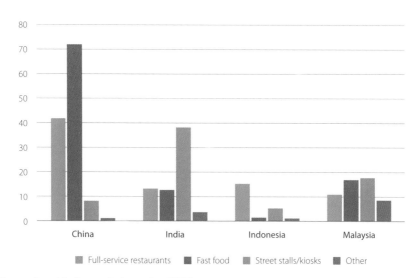

Source of raw data: Euromonitor International (2017)
Note: "Other" includes home delivery/takeaway, cafes/bars and self-service cafeterias.

The types of vendors vary not only across countries, but also across different sized cities. Furthermore, the availability of different types of vendors can affect the types of food people eat. For example, in Indonesia, it was found that infant feeding patterns were different in urban slum areas of Jakarta versus peri-urban slum areas surrounding the city of Yogyakarta (Martha, Amelia and Myranti, 2017). While some practices by caregivers are common in both areas (such as feeding rice with vegetable soup), in urban slum areas caregivers have wider access to street food vendors and more frequently purchase snack foods such as fried tofu, tempeh (soybean cake), bananas and manufactured biscuits. Caregivers in these areas are also more likely to buy side dishes and vegetables cooked in nearby street stalls because their houses are small and less likely to be equipped with a kitchen. Caregivers in both areas perceive home-cooked meals as healthier, but structural issues lead poor urban caregivers to cook less. One implication is that generic education campaigns may not change feeding practices unless they are tailored to the location-specific constraints faced by households.

The increased importance of eating food prepared away from home, at least in parts of the region, has several important implications. It has implications for nutrition, as consumers may be less aware of the ingredients added to their food. Therefore, interventions to raise awareness might need to target food service providers as well as consumers. It has implications for food safety, as it is complex to regulate the multiplicity of restaurants and street vendors. It also has important implications for data collection and evidence-based policy-making. Household income and expenditure surveys typically record the total amount of expenditures on food prepared outside the home, but not the types of food consumed. The fact that such a large share of food is prepared outside the home makes such surveys much less valuable in understanding food consumption patterns across different groups of households, and more sophisticated surveys will need to be conducted in the future if we are to understand consumption patterns better.

Increasing diversity of retail market outlets

While expenditures on food prepared away from home are rising, most food expenditures are still for food prepared at home. In this regard, consumers increasingly purchase their food from a wide range of markets and outlets, including modern retail sources (hypermarkets, supermarkets, convenience stores), traditional markets and street vendors. For example, residents of Dhaka might purchase staple foods at the neighbourhood grocery store for the sake of convenience, packaged processed food at a supermarket due to greater availability and choice, and fresh fish at a wet market in order to obtain the lowest price (FAO, 2018a). The wide range of retail outlets implies that policies will need to influence a variety of food retail service outlets to ensure food safety and encourage the purchase of nutritious foods.

While supermarkets have grown rapidly (Reardon *et al.*, 2003), especially in China, there is nevertheless a wide diversity of market types competing for different types of consumers and meeting a variety of demands At national levels, traditional retail channels (e.g. wet markets, street vendors, small shops) still account for the dominant share (more than 80 percent) of the food retail market in Indonesia and India, and a greater than 50 percent share in Malaysia (Euromonitor International, 2017).

In Indonesia (Minot *et al.,* 2015), found that greater income, more education and ownership of a vehicle or refrigerator all increased the share of the food budget spent in modern retail. Younger people also tended to spend a greater share of their food budgets at modern retail establishments, although families with working wives did not (after controlling for other factors). The types of food sold at modern retail establishments in Indonesia were primarily dairy products, processed foods and imported fruit (e.g. apples), while vegetables and domestically produced fruit were sold almost exclusively through traditional retail outlets. In China, about 70 percent of urban consumers obtain their fruits and vegetables from traditional markets (Zhang and Pan, 2013).

The fact that traditional retail remains a major source of food for urban consumers suggests that the impact of supermarket quality standards on smallholder farmers may be less than feared by many, at least for now. In addition, the dominant share of traditional wet markets, itinerant vendors and small shops in sales of meat, fish and vegetables suggests that these food products are widely available in urban areas (Minot *et al.,* 2015).

Food manufacturing

The demands of consumers for more convenience, improved food safety, higher quality and other characteristics lead to the growth of an agribusiness sector that makes investments in food processing and distribution to deliver a variety of food types in different locations. This agribusiness sector grows in importance relative to primary agriculture as GDP per capita increases (Figure 2.16). The investments in food processing and distribution raise the price of the final product and thus lower the share of the farm in final value added (see Figure 5.8 for an example concerning dairy products in the region). This pattern is a natural consequence of the fact that more processing takes place after the product leaves the farm and has also occurred in high-income countries (Antle, 1999).

The decreasing share to farmers does not imply that farmers lose out, as a smaller share of a larger amount is not equivalent to a smaller absolute amount. In fact, the additional processing can create new markets for farmers that allow them to increase profits, especially if the agribusiness firms are located in small- or medium-sized cities with closer ties to rural areas (FAO, 2017). In addition to new output markets, the additional processing can also strengthen links between these processing firms and farmers that help farmers to gain access to better seeds and breeds and new technologies, thereby improving productivity (see next section).

Figure 5.8 Percentage of value added post farm gate relative to consumer price of milk, various countries

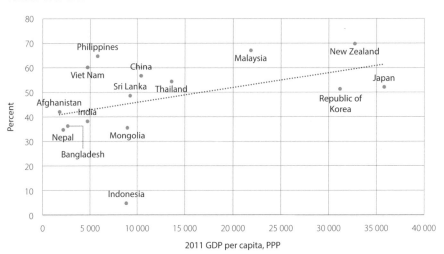

Source: IFCN (2012)

Food processing firms can also be a source of employment. In China, India and Indonesia, the vast bulk of food processing firms are very small firms employing fewer than ten people, especially in the latter two countries where such firms account for more than 98 percent of the total. However, these small enterprises account for a small percentage of sales, less than 1 percent in India and Indonesia and 2–3 percent in China. Firms with more than 50 employees, while small in number, account for 60–80 percent of total sales in these countries (all data in this paragraph come from Euromonitor International, 2017). This pattern may be due to economies of scale or reputational issues related to food safety.

Moving inputs from factories to farm, and food from farm to retail

Farm production depends on availability of and access to a number of inputs (e.g. seeds, fertilizer, animal feed). Thus, value chains that deliver food to consumers do not begin at the farm – inputs must first be delivered to farms for the production process to begin. These value chains are becoming steadily more sophisticated in the region, what has been termed a 'quiet revolution' by Reardon and others (Reardon *et al.*, 2012, 2014).

There are various models to move food from farm to retail (and inputs to farms). Traditional 'decentralized' models rely on individual aggregators, wholesale markets and spot market transactions to supply retail vendors, and such systems are still dominant in the region for the marketing of staple foods, as well as for fruits and vegetables, other foods and production inputs.

For example, the expansion of the aquaculture sector in Myanmar has been facilitated by rapid increases in the number of private sector fingerling suppliers who invest in hatcheries, nurseries and boats to deliver seed to fishpond operators. Feed supply in this value chain is also important, although this has been constrained in Myanmar by a lack of competition that keeps pelleted feed prices high (Belton *et al.*, 2018). In Andhra Pradesh (India), however, there has been rapid investment in feed mills to serve the growing aquaculture industry, nearly all of it funded by domestic entrepreneurs (Belton *et al.*, 2017).

Value chains for other commodities have also expanded. For example, in India, increased government investment in electrification and roads, as well as subsidies for the construction and expansion of cold chains, have spurred the private sector to develop an improved network of potato cold storage facilities that allow storage for seven months instead of three. Within the span of a decade, the share of potato farmers in Agra District near Delhi using cold storage more than doubled, from 40 to 95 percent. The increased ease of storage has also allowed farmers greater choice of buyers for their produce (Dasgupta *et al.*, 2010; FAO, 2017; Reardon *et al.*, 2012).

The role of the private sector in the development of these value chains is essential, as the scale of investment and the local knowledge required are far beyond the capacity of any government to provide. Even in countries such as India where the government tends to play a larger role than in other countries, the government's role as a direct buyer and seller accounts for less than 10 percent of India's total food economy (Reardon and Minten, 2011).

While 'decentralized' models like those noted above are still common, increased vertical coordination is emerging in the region, particularly in the dairy and livestock sectors (Verhofstadt, Maertens and Swinnen, 2013). Among other models, coordination with farmers can involve downstream buyers of production, upstream suppliers of inputs (e.g. animals, feed), fellow farmers (e.g. cooperatives), production complexes where farmers work together with other farmers and technical staff in a designated area or multi-stakeholder (traders, processors, local government and research institutes) cluster models. Another model is complete vertical integration where one firm rents or owns the farm and owns and manages all of the subsequent processing and distribution. However, this model is less common in Asia and the Pacific than in other regions. A driving force behind all of these coordination models is an effort to improve quality and food safety (Verhofstadt, Maertens and Swinnen, 2013).

As demands for safer, higher quality food increase, a major challenge in these emerging value chains will be to make sure they are inclusive of smallholder farm households through various mechanisms such as contract farming (Miyata, Minot and Hu, 2009). One way to make growth more inclusive is to include more farmers in such value chains, although there may be a bias towards including larger farms owned by relatively well-educated farmers, as this reduces transaction costs. Wage employment in large-scale production offers another alternative that can improve rural standards of living, and should not be overlooked as a way to make rural development more inclusive (Maertens, Colen and Swinnen, 2011; Maertens and Swinnen, 2009). It can be especially beneficial to the rural landless who have no opportunity to sell products into value chains because they have no land other than for their houses and a small garden, or to women who may be excluded from participating in product marketing chains (Maertens, Minten and Swinnen, 2012).

International trade and globalization

As noted in Chapter 2, international trade in food grew rapidly for most countries in the region after 2000 due to improved infrastructure, innovations in logistics and free trade agreements. While most countries produce the bulk of their dietary energy supplies domestically (Map 2.3), international trade provides the option for consumers to source a wider variety of foods at lower prices, making these foods more affordable to consumers. For example, apples are now more widely available in tropical countries due to the growth in trade. In general, international trade can improve or worsen the healthiness of diets, depending upon what foods are traded and in what amounts (e.g. greater trade in fruits and vegetables versus greater trade in fatty parts of animals, as has occurred in the Pacific Islands).

Greater international trade also affects the livelihoods of farmers, giving some of them a chance to export to new markets, but also opening others to increased competition from abroad. These impacts can be moderated by trade policy, and different countries are pursuing different strategies in this regard, as illustrated by a comparison between Viet Nam and Indonesia.

Since the turn of the century, Viet Nam has been exporting increasingly large quantities of various different foods, including rice, fish, fruits and vegetables (Figure 5.9a). At the same time, it has also been importing increasingly large quantities of other foods, such as wheat, maize and oilseeds for animal feed, and dairy products. These trends suggest a dynamic economy that is pursuing its comparative advantage in the production of certain types of food and importing other types of food. Indonesia, in contrast (Figure 5.9b), has seen much smaller changes in trade, as it has placed more policy emphasis on self-sufficiency in a wide range of commodities, including rice, maize, soybean, beef and sugar (Sudaryanto, 2016) and has erected other barriers to trade in fruits and vegetables (World Bank, 2016). Such trade barriers, if perceived as temporary, can encourage farmers to innovate and become more competitive, or give governments time to provide infrastructure that promotes competitiveness. However, these barriers also make food more expensive, making it more difficult for the poor to afford nutritious food (Block *et al.*, 2004) and phasing out such barriers over time is often politically challenging. Thus, different countries have different perspectives on the role of international trade in improving farmer competitiveness and reducing malnutrition.

Figure 5.9a International trade in various food items, Viet Nam, 2001–2003 and 2014–2016

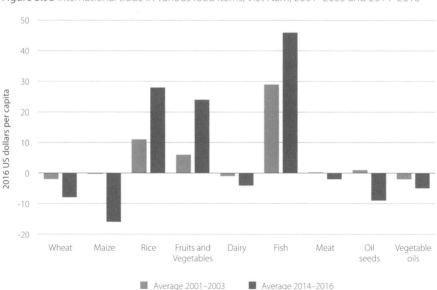

Figure 5.9b International trade in various food items, Indonesia, 2001–2003 and 2014–2016

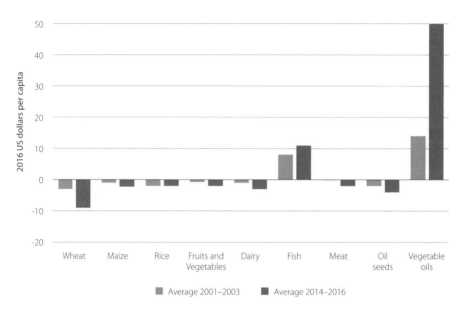

Sources of raw data: FAO (2018), IMF (2018), ITC (2018)
Note: Vertical axis truncated at 50 US dollars per capita in order to show trends in other commodities more clearly. Value for vegetable oils for Indonesia in 2014–2016 is $75.

Summary

Transport and communications infrastructures have expanded rapidly in most of the region, although a large percentage of the population does not have access to the Internet and electrification has not reached all rural people, especially in South Asia and the Pacific. These infrastructure improvements have supported the delivery of improved inputs to farmers, increases in food manufacturing to cater to changing food demands and movement of food to urban areas in response to rapid urbanization. These value chains have also expanded across international borders, facilitated by better infrastructure and trade agreements.

People living in urban areas face more traffic congestion, live a longer distance from the workplace and have more rigid work schedules and longer working hours relative to rural areas. As a result, urbanization has led to increased demand for more convenient food preparation, which has driven a transformation of urban food environments, in particular an increase in the diversity of retail markets outlets and establishments for eating food prepared away from home.

References

Antle, J.M. 1999. The new economics of agriculture. *American Journal of Agricultural Economics*, 81(5): 993–1010. https://doi.org/10.2307/1244078

Belton, B., Hein, A., Htoo, K., Kham, L.S., Phyoe, A.S. & Reardon, T. 2018. The emerging quiet revolution in Myanmar's aquaculture value chain. *Aquaculture*, 493: 384–394. https://doi.org/10.1016/j.aquaculture.2017.06.028

Belton, B., Padiyar, A., G, R. & K, G.R. 2017. Boom and bust in Andhra Pradesh: Development and transformation in India's domestic aquaculture value chain. *Aquaculture*, 470: 196–206. https://doi.org/10.1016/j.aquaculture.2016.12.019

Block, S.A., Kiess, L., Webb, P., Kosen, S., Moench-Pfanner, R., Bloem, M. & Timmer, P. 2004. Macro shocks and micro outcomes: child nutrition during Indonesia's crisis. Economics and Human Biology, 2: 21–44. https://doi.org/10.1016/j.ehb.2003.12.007

Dasgupta, S., Reardon, T., Minten, B. & Singh, S. 2010. The transforming potato value chain in India: Potato pathways from a commercialized-agriculture zone (Agra) to Delhi. New Delhi, ADB-IFPRI.

Euromonitor International. 2017. *Passport global market information database* [online]. http://www.portal.euromonitor.com

FAO. 2017. *The state of food and agriculture: Leveraging food systems for inclusive rural transformation*. Rome. (also available at http://www.fao.org/publications/sofa/en/).

FAO. 2018a. *Asia and the Pacific regional overview of food security and nutrition*. Bangkok, FAO Regional Office for Asia and the Pacific.

FAO. 2018b. *FAOSTAT* [online]. www.fao.org/faostat/

IMF. 2018. *International Financial Statistics (IFS) Data* [online]. http://www.imf.org/en/Data

International Farm Comparison Network (IFCN). 2012. Dairy report 2012: For a better understanding of milk production world-wide. (also available at https://ifcndairy.org/ifcn-products-services/dairy-report/).

International Trade Center (ITC). 2018. *ITC trade map* [online]. https://www.trademap.org/Index.aspx

Maertens, M., Colen, L. & Swinnen, J.F.M. 2011. Globalisation and poverty in Senegal: a worst case scenario? *European Review of Agricultural Economics*, 38(1): 31–54. https://doi.org/10.1093/erae/jbq053

Maertens, M., Minten, B. & Swinnen, J. 2012. Modern food supply chains and development: Evidence from horticulture export sectors in Sub-Saharan Africa. *Development Policy Review*, 30(4): 473–497. https://doi.org/10.1111/j.1467-7679.2012.00585.x

Maertens, M. & Swinnen, J.F.M. 2009. Trade, standards, and poverty: Evidence from Senegal. *World Development*, 37(1): 161–178. https://doi.org/10.1016/j.worlddev.2008.04.006

Martha, E., Amelia, T. & Myranti. 2017. Toddler's eating behaviour in slum urban and semi urban communities: Study in Kampung Melayu and Bantul, Indonesia. *The 1st International Conference on Global Health, 9–11 November 2017, Jakarta, Indonesia*: 1–7. https://doi.org/10.18502/kls.v4i1.1360

Minot, N., Stringer, R., Umberger, W.J. & Maghraby, W. 2015. Urban shopping patterns in Indonesia and their implications for small farmers. *Bulletin of Indonesian Economic Studies*, 51(3): 375–388. https://doi.org/10.1080/00074918.2015.1104410

Miyata, S., Minot, N. & Hu, D. 2009. Impact of contract farming on income: Linking small farmers, packers, and supermarkets in China. *World Development*, 37(11): 1781–1790. https://doi.org/10.1016/j.worlddev.2008.08.025

Reardon, T., Chen, K., Minten, B. & Adriano, L. 2012. *The quiet revolution in staple food value chains: Enter the dragon, the elephant and the tiger.* ADB & IFPRI. (also available at http://www.ifpri.org/publication/quiet-revolution-staple-food-value-chains).

Reardon, T. & Minten, B. 2011. The quiet revolution in India's food supply chains. *IFPRI Discussion Paper 01115*. (also available at http://www.ifpri.org/publication/quiet-revolution-indias-food-supply-chains-0).

Reardon, T., Timmer, C.P., Barrett, C.B. & Berdegué, J. 2003. The rise of supermarkets in Africa, Asia, and Latin America. *American Journal of Agricultural Economics*, 85(5): 1140–1146.https://doi.org/10.1111/j.0092-5853.2003.00520.x

Reardon, T., Tschirley, D., Dolislager, M., Snyder, J., Hu, C. & White, S. 2014. Urbanization, diet change, and transformation of food supply chains in Asia. (May). (also available at http://www.fao.org/fileadmin/templates/ags/docs/MUFN/DOCUMENTS/MUS_Reardon_2014.pdf).

Sahakian, M.; Saloma, C.; Erkman, S.. eds. 2017. *Food consumption in the city: practices and patterns in urban Asia and the Pacific.* London, Routledge.

Sudaryanto, T. 2016. Government policy on self sufficiency to achieve food security in Indonesia. *FFTC Agricultural Policy Articles.* (also available at http://ap.fftc.agnet.org/ap_db.php?id=680&print=1).

The Economic and Social Commission for Asia and the Pacific (UNESCAP). 2018. *ESCAP Online Statistical Database* [online]. [Cited 20 June 2018]. http://data.unescap.org/escap_stat/#data/

Verhofstadt, E., Maertens, M. & Swinnen, J. 2013. Smallholder participation in transforming agri-food supply chains in East Asia. Farmgate-to-market study on managing the agri-food transition in East Asia No. 3

Warr, P., Widodo, M.A. & Yusuf, A.A. 2018. Urbanization and the demand for food in Indonesia

World Bank. 2016. *World development indicators 2016.* (also available at http://hdl.handle.net/10986/23969).

World Bank. 2018. *World development indicators* [online]. https://data.worldbank.org/products/wdi

Zhang, Q.F. & Pan, Z. 2013. The transformation of urban vegetable retail in China: Wet markets, supermarkets and informal markets in Shanghai. *Journal of Contemporary Asia*, 43(3): 497–518. https://doi.org/10.1080/00472336.2013.782224

RURAL LIVELIHOODS AND LABOUR PRODUCTIVITY IN FARMING SYSTEMS

Farming systems have several important functions. They should produce nutritious food efficiently and competitively so that it is affordable to poor consumers in both urban and rural areas and can contribute to the eradication of malnutrition (see Chapter 4). They should produce food sustainably so as to enhance the environment and not compromise future food production capacity (see Chapter 3). Equally important, in the context of the overall rural economy, farming systems must provide a decent standard of living for rural families, including the smallholders who comprise the vast bulk of farm families in the region (see Chapter 1) as well as the poor rural landless who own little or no land.

Trends in rural livelihoods and demographics

Declining rural poverty

As with overall poverty rates, rural poverty is falling substantially throughout the region in nearly all countries (Figure 6.1), although progress in the Pacific may be more problematic (Fiji is the only Pacific country with data for two points in time). The falling rural poverty in Asia is due to rapid economic growth that has reached the poor (Chapter 2), even if inequality remains an issue.

Figure 6.1 Rural poverty rate, selected countries 1990–2015

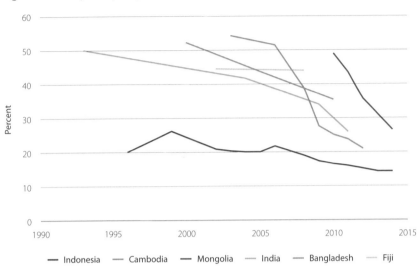

Source of raw data: World Bank (2018)

Growing importance of rural non-farm income

In the context of the economic growth described in Chapter 2, agriculture's relative share of the overall economy is declining (structural transformation), even as the agriculture sector grows in absolute terms. While urbanization is an important part of this process, the growing importance of non-farm income is not confined to urban areas, but reaches well into the countryside. Indeed, rural economies in the region have been diversified for longer than is often realized – people have never lived by bread (rice) alone. In 1840s Tokugawa Japan, income from non-agricultural sources was usually more than half of total household income, despite the fact that most families considered their occupation to be farming. As one village headman stated at the time: "My father always said that the people of this village could not prosper either by farming or by commerce alone but must carry on both together – just as a cart must have two wheels" (Smith, 1988). And, just as today, many observers at that time complained about people neglecting farming for other income-generating activities, even though participation in non-farm activities was an important pathway for rural families to increase per capita household income.

Non-farm income remains of paramount importance to farm families today. Data from various household surveys described in Davis, Di Giuseppe and Zezza (2017) show that nearly all rural households in a selection of Asian countries (Bangladesh, Indonesia, Nepal, Pakistan and Viet Nam) earned income from non-farm activities (Figure 6.2).[1] Of course, many households are also simultaneously engaged in agriculture, but the share of households that are specialized in agriculture is declining over time in four out of these five countries (the green bars in Figure 6.3). The only exception is Bangladesh, where the share of households specialized in farming is very low in any event. Examples of rural non-farm income include sewing and weaving, construction, teaching, provision of medical care, trading and local transportation. Many households also earn non-farm income from members who permanently or temporarily move to urban areas.

Figure 6.2 Percentage of rural households earning income from non-farm activities, early 2000s

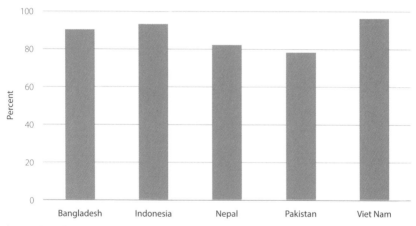

Source of raw data: Davis *et al.* (2017)

Note: Specific years for each country are as follows: Bangladesh 2005, Indonesia 2000, Nepal 2003, Pakistan 2001 and Viet Nam 2002.

1   Despite the title of the cited paper, data for Asia are included.

Figure 6.3 Livelihood patterns in rural households over time

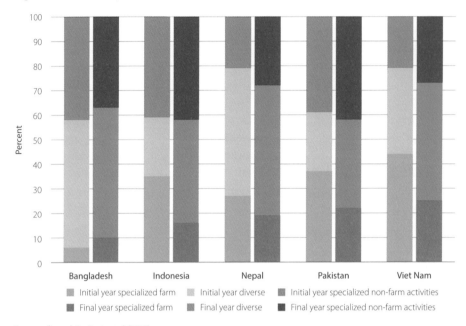

Source of raw data: Davis *et al.* (2017)

Note: Households are considered specialized in farming or non-farm activities if 75 percent or more of their income comes from that sector.

While the share of households specialized in farming is declining, the share of households that are specialized in non-farm income activities (the blue bars) is generally increasing, as is the share of households that earn income from a diverse range of activities (with no single sector contributing more than 75 percent of household income).

The growing importance of non-farm enterprises in rural areas has both advantages and disadvantages. In terms of advantages, it increases labour demand and thus wages and employment in rural areas, both of which contribute to increased household income. Increased labour demand, especially for unskilled labour, benefits the poor because labour is typically their main or only asset. The increased non-farm demand for labour is also helpful in the context of agriculture's seasonality by providing employment outside of planting and harvesting seasons.

The main reason why people are keeping one foot in agriculture and one foot in the city is that most countries lack a comprehensive system of social safety nets. Keeping some agricultural land serves as a safety net in case of crisis: for example, millions of Indonesians, having lost urban employment during the Asian financial crisis in 1998, migrated back to rural areas (Bresciani *et al.,* 2002). Growth of the rural non-farm economy also allows some people to exit agriculture without leaving rural areas.

Rural non-farm income also helps to diversify the structure of household income, which makes it easier for households to manage the risks inherent in livelihoods that are tied to agriculture, with its dependence on short-term fluctuations in the weather. Non-farm income can also compensate for poorly functioning credit markets by giving farmers an alternative means of financing investments in agricultural inputs (Reardon *et al.,* 2012, 2014).

While additional rural non-farm income brings a wealth of benefits and opportunities to rural households, it also creates some challenges for the agriculture sector by potentially affecting the adoption of new innovative technologies that can increase profits and improve the environment but require time to learn and adopt. Learning a new technology or management technique has a large element of fixed costs that is independent of farm size (Dawe, 2003; Foster and Rosenzweig, 2017), which implies that small farms will have less incentive to adopt such technologies because they can be utilized only over a small area. By increasing the opportunity cost of time, especially in the context of small and declining farm sizes in the region (see below), the growing importance of non-farm income can reduce farmers' incentives to spend the time required to learn new technologies. In this environment, institutional innovations may be necessary to make the agriculture sector efficient and competitive (see Chapter 7).

Ageing of rural areas

As discussed in Chapter 2, urban areas have a greater proportion of people of prime working age (15–49 years) than rural areas, as young people increasingly move to cities in search of employment. As a result, the average age of farmers is increasing, especially in developed countries. In the Republic of Korea, the average age of farm operators increased from 56 to 66 between 1995 and 2015. The age distribution is quite striking, with farmers over the age of 70 now being the largest single group (Figure 6.4). Farm operators over the age of 60 account for 68 percent of all farmers, with those under 50 accounting for just 9 percent (Lee, 2017). In Japan, the average age of farmers increased from 59 to 66 during the same period.

Figure 6.4 Distribution of farmers by age group, Republic of Korea, 2015

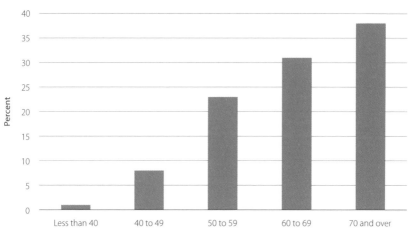

Source of raw data: Lee (2017)

But this phenomenon is not only confined to OECD economies – similar trends are taking hold elsewhere as well. In Viet Nam, the mean age for all those working in agriculture as their primary economic activity increased from 33 to 41 between 2001 and 2011 (Viet Nam Agricultural Census). In Bangladesh, the average age of the farm household head increased from 35 to 48 between 1988 and 2014 (Bhandari and Mishra, 2018). In these countries, there is less of an immediate problem relative to that in wealthier economies, as farmers in their 40s may have a good mix of strength and experience.

Nevertheless, the trend for the future is clear. There will be increased labour shortages in farm households, as older people are less able to carry out physically demanding farm work. Older farmers may also be less receptive to the adoption of new technologies, as they have fewer years remaining in the labour force when they can use them. At the same time, increasing land scarcity and in some cases fragmentation (see the section on farm sizes below) make it difficult for young people to gain access to land, which has been identified as one of the main factors in young people's lack of interest in farming (Li, 2017; Park and White, 2017; Portilla, 2017; White, 2018).

Feminization of agricultural employment

As discussed in Chapter 2, urban areas in South Asia have a higher share of men than rural areas, although this is not the case for the rest of the region. In fact, in East Asia, Southeast Asia and the Pacific the tendency is for urban areas in most countries to have a higher share of women than rural areas. The different pattern in South Asia may be due to different cultural norms that are reflected in different degrees of women's empowerment and constraints on being outside the home (as reflected in the lower value of the Gender Development Index for South Asia noted in Chapter 2). Consistent with these differential gender patterns of migration, there has been increased feminization of the agricultural labour force in South Asia, particularly in Nepal (Figure 6.5). In East and Southeast Asia, however, the agricultural labour force is becoming more 'masculinized' over time, especially in the high-income countries (Figure 6.6).

Figure 6.5 Ratio of female to male employment in agriculture in South Asia

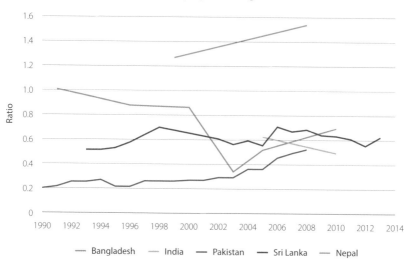

Source of raw data: World Bank (2018)

Figure 6.6 Ratio of female to male employment in agriculture in East and Southeast Asia

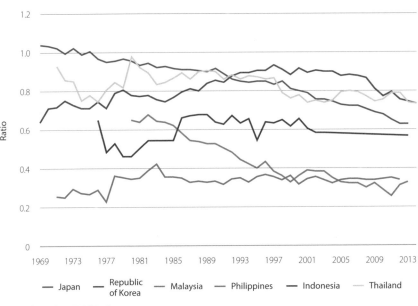

Source of raw data: FAO (2018)

Gender roles within agriculture are more complex than indicated by the above data, however. For some farm operations, women and men work together at the same time, e.g. for rice harvesting. However, in other cases, one sex or the other dominates a particular task. For example, for rice crop establishment and weeding in Tamil Nadu, women perform the bulk of the work. On the other hand, men carry out nearly all of the land preparation activities for the rice crop in a range of different countries around Asia (spanning the different subregions). The gender distribution of labour has implications for the social impacts of mechanization, some of which are discussed later in this chapter. Women also often receive lower wages for identical tasks (Paris, 2017).

Trends in farming systems, causes and implications

Greater labour productivity

For low-income countries, the share of the labour force working in agriculture is substantially greater than the share of agriculture in GDP (e.g. agriculture accounts for about 44 percent of employment in South Asia compared to just 18 percent of value added (see Figures 2.14 and 2.15). This state of affairs reflects an intersectoral productivity gap – labour productivity (value added per worker) is lower in agriculture than in the rest of the economy. In very high-income countries, this gap tends to shrink substantially, if not disappear altogether (in high-income countries in the region, agriculture accounts for about 4 percent of employment and 2 percent of GDP). The process of increasing labour productivity in agriculture in particular (and in rural areas in general) is critical to creating a more inclusive society where those employed in agriculture earn comparable incomes to those outside agriculture because household income is closely tied to labour productivity.

Mathematically, shrinking the intersectoral gap in labour productivity can either come from growth in agricultural value added that exceeds growth in manufacturing and services, or by slower growth in employment in agriculture (again, relative to manufacturing and services). While agricultural value added has grown considerably in all subregions other than the Pacific, consumer demand drives rapid growth in manufacturing and services that consistently exceeds growth in agriculture (Figure 2.13). Thus, agricultural growth needs to be supplemented with relatively slow (or negative) growth of employment in agriculture to help reduce the intersectoral gap in labour productivity. In fact, from 1991 to 2017, the most rapid increases in agricultural value added per worker (labour productivity) in the region occurred in East Asia and among the high-income countries, where large declines in the number of agricultural workers were responsible for about half of the increase in agricultural value added per worker (Figure 6.7).

Figure 6.7 Average annual growth in agricultural value-added (VA) and agricultural VA per worker, 1991–2016

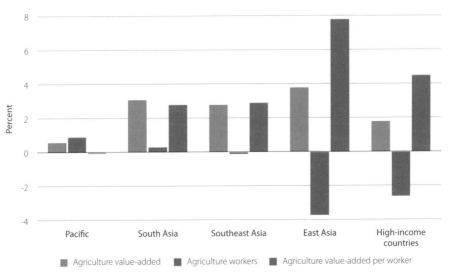

■ Agriculture value-added   ■ Agriculture workers   ■ Agriculture value-added per worker

Source of raw data: World Bank (2018)

Note: Average annual growth rates calculated as slope of regression line over the time period.

During the past 25 years, the number of agricultural workers has declined more rapidly in countries with higher levels of GDP per capita. Thus, among the four developing subregions, there was a large decline in East Asia during the past 25 years, a small decline in Southeast Asia, a small increase in South Asia and a larger increase in the Pacific.

The process of labour leaving agriculture, and capital entering the sector to support mechanization (see next section), leads to higher capital-to-labour ratios and higher labour productivity for those who remain in the sector. While exit of labour from agriculture will be the dominant force over the longer term, there are other ways to increase labour productivity.

For example, in addition to reducing labour input, it is also possible to increase value added per worker in other ways. One option is a higher price for output, for example by diversifying to higher value food products. A second option is to increase yields, which raises the value of output. Yet another option is to adopt environmentally friendly production methods and reduce material input costs, provided there is not a large decline in yields.

The exit of labour from agriculture in macro terms is not to deny the fact that many farmers will remain in the sector for many years to come, especially in poorer countries. But if these farmers are to share fully in economy-wide income growth, they will need to find very high-value markets that can support full-time employment in agriculture. These opportunities exist, especially for high-value foods and organic production (see the last section in Chapter 4), but there are not enough of those opportunities to negate the macro long-term exit trend from agriculture. Alternatively, farm households can diversify into the non-farm sector while keeping one foot in agriculture, as described earlier in this chapter.

Rising rural wages and mechanization

The exit of labour from agriculture for jobs in cities or overseas (or due to old age) has an impact on the sector – it tends to increase labour scarcity, push up rural wages and encourage mechanization.[2] Data on real rural wages are scarce, but rising rural wages have been documented in a number of Asian countries (Wiggins and Keats, 2014). Data from the Pacific, where agricultural growth has been slow (Figure 2.13), are less clear.

Mechanization and labour saving innovations in food systems have a long history in the region. Harvesting practices have evolved from the use of small knives (e.g. ani-ani in Indonesia) to sickles and are now moving to combine-harvesters today. Hand pounding of rice has given way to small portable mills (that are much bigger today), herbicides have replaced much hand weeding, motors have replaced treadle pumps for irrigation and broadcast seeding has replaced laborious manual transplanting as a method of crop establishment (with mechanical transplanters also set to replace broadcast seeding in some areas).

Further mechanization is taking place throughout the region. In India, tractor sales tripled in the decade from 2003/2004 to 2013/2014, and the number of farms using tractors nearly doubled in just five years, from 2006/2007 to 2011/2012 (Gulati, Saini and Manchanda, 2017). Rice farmers in the Mekong Delta and Zhejiang Province in China have reduced labour use substantially since the turn of the century by mechanization of harvest operations and switching to direct seeding (Dawe, 2015). Mechanical rice transplanting is gaining ground in Thailand (Poapongsakorn, Pantakua and Wiwatvicha, 2016). Mechanization is also spreading rapidly in Myanmar, where prices for machines are falling as wages are rising (Cho, 2017), and in Bangladesh (Box 3).

---

2   Mechanization/automation can also occur when real wages are not rising. But rising real wages increase the incentives for mechanization of farm operations.

Box 3 Mechanization in northwestern Bangladesh increases cropping intensity and improves natural resource management

Mechanization is not only a labour-saving strategy, but it also can allow for an increase in cropping intensity (the number of crops planted per year) and improved utilization of scarce natural resources. In Rajshahi (northwestern Bangladesh), entrepreneurs operate two wheel tractors that perform conservation tillage, seeding and fertilization in a single pass through the field. This combination of operations allows farmers to overcome seasonal labour bottlenecks and seed the crop in a timely manner that allows for utilization of residual soil moisture from the previous rainy season crop, thus improving water productivity and overcoming drought conditions. It increases land productivity because it reduces the turn-around time between crops, thereby allowing triple cropping that includes a crop of pulses such as chickpea or lentils. Reduced labour inputs lead to lower production costs, about USD 23 per hectare for crop establishment instead of USD 80 per hectare.

Some challenges encountered by service providers include a shortage of trained machine operators in the area and a lack of spare parts. Seeding depth can sometimes be irregular, and machine operators must spend many hours in a standing position when operating two wheel tractors.

Most machinery is relatively expensive compared to other farm inputs and this expense might present a significant obstacle to adoption. In practice, however, there are active machinery rental markets for a wide range of services that have allowed many farmers to mechanize without owning a machine. In the Philippines, mechanical threshing of the rice crop became widespread in the mid-1970s, as a small number of farmer-entrepreneurs bought the new axial-flow threshers and moved them around the countryside to serve anyone who wanted those services. Because it was cheaper than hiring labour to do the same task, this technology spread rapidly. In the Central Plains of Thailand, combine-harvesting is widespread, even though only 3 percent of farmers own such a machine. Individual entrepreneurs operate most of these services, with cooperatives supplying only a small percentage (Dawe, 2005). In Bangladesh, while 72 percent of farmers use power tillers, only 2 percent of farmers own them (Mandal, 2014).

Mechanization services are not limited to land preparation and harvesting. Once Bangladesh lowered tariffs on imports of low-lift water pumps in the 1980s (Hossain, 1996), imports of such equipment surged and farmers rented them out to their neighbours, who were then able to irrigate their farms. In Indonesia and the Philippines, many mango farmers rent spraying services to regulate flowering (Qanti, Reardon and Iswariyadi, 2017).

Mechanization generally lowers production costs – for example, costs of rice production per kilogram are much lower in the production areas where there is mechanical harvesting and manual transplanting has disappeared (Moya *et al.*, 2016). Mechanization also opens up opportunities for diversification. In India, for example, with the spread of tractors, the main purpose of cattle has shifted from ploughing fields and local transport to milk production (Kishore *et al.*, 2016).

Like most changes, however, mechanization has advantages and disadvantages, including some negative social implications, as the reduction in labour use causes people to lose their jobs. Job loss is a painful transition for anyone, especially if the person is old or middle-aged and has difficulty finding alternative employment but is not able to retire. There is also often a gender dimension to this transition, as operation of machines is typically a male domain (FAO, 2015; FAO, 2016). If a farm operation that employs women is mechanized, then women most likely suffer disproportionately from the job loss. To the extent that women are overworked and time-poor, the reduction in workload and drudgery (due for example to adoption of water pumps) may be welcome, but it may also cause women to lose control over a source of income, which could in turn affect their empowerment and the nutritional status of children in the household (Allendorf, 2007; Rahman, Saima and Goni, 2015). To the extent that people lose their jobs and cannot find employment in the emerging non-farm sector or elsewhere in the agriculture sector (e.g. in the production of other foods, see section below), social protection has an important function, especially if the extended family is unable to provide support.

Changes in farm size and operation

In addition to mechanization, the rising demand for labour in urban areas might also lead to larger farm sizes, as workers exit farming and leave a shrinking pool of labour to handle ever-larger farms. There are two constraints to farm sizes becoming larger, however. First, land in most of the region is scarce and the quantity of arable land available is shrinking due to urbanization, road construction and other demands. Second, and more important, the choice between earning agricultural and non-agricultural income is not a binary (i.e. either, or): many people are maintaining their ties to agriculture while gradually increasing the importance of non-farm employment in total household income (see earlier section in this chapter). In some countries (e.g. the Philippines), there have also been agrarian reform programmes that have broken up large farms into large numbers of smaller farms (Borras, 2005; Deininger, Olinto and Maertens, 2000).

Given these considerations, farm size has been declining in nearly all countries in South Asia, Southeast Asia and the Pacific over the past few decades (Figure 6.8a, Figure 6.8b, Figure 6.8d).[3] In East Asia, farm sizes have started to increase (Figure 6.8c), but the absolute magnitude of these increases has been very small. Given the general scarcity of land in Asia and the Pacific relative to other continents (Figure 1.3), it is evident that farm sizes in the region will never be as large as those in the rest of the world.

Figure 6.8a Trends in farm size in South Asia, 1960–2010

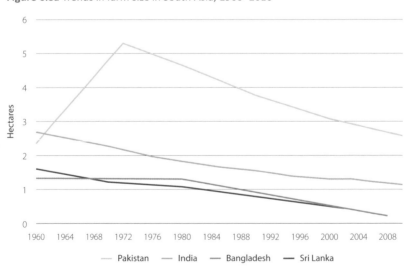

---

3 Agricultural censuses define farm size as the farm area operated, not owned. Thus, these data take into account the fact that some farmers rent in (and rent out) land.

Figure 6.8b Trends in farm size in Southeast Asia, 1960–2013

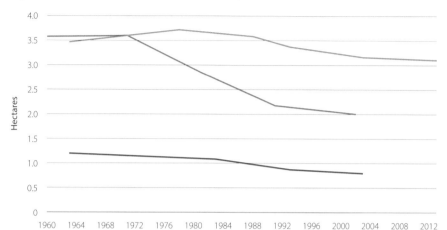

Figure 6.8c Trends in farm size in East Asia, 1960–2013

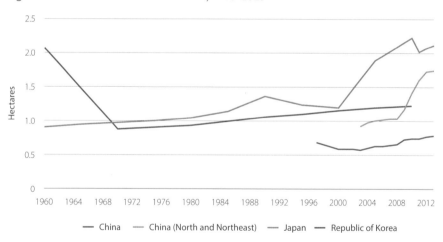

Figure 6.8d Trends in farm size in the Pacific, 1970–2015

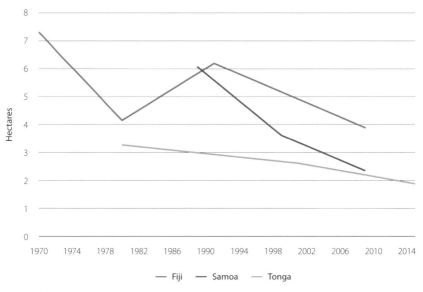

Source of raw data: FAO (2018b)

It is important to note that while average farm sizes are declining in much of the region, there are some notable trends towards consolidation in specific circumstances, one example being the livestock sector. Pig farming is becoming larger scale in the region and the same trend is taking place for poultry and dairy farming, at least in some countries (notably China and Thailand) (Poapongsakorn, Pantakua and Wiwatvicha, 2016; Verhofstadt, Maertens and Swinnen, 2013). Consolidation can have advantages in terms of unit production costs and traceability, but can also have serious animal welfare and environmental drawbacks due to the concentration of large numbers of animals (and thus, quantities of manure) in a relatively small space.

Mathematically, farm sizes will decline if the percentage change in total farm area is less than the percentage change in the number of farms. Given the scarcity of land in the region, total farm area is not increasing rapidly. Meanwhile, despite continued rural to urban migration, the number of farms is not decreasing very rapidly as rural populations continue to increase in poorer countries (and farmers subdivide their land). Even in wealthier countries where the rural population is declining, urban households often maintain their farms as a safety net in times of crisis, as discussed earlier. The combined effect of these factors is that farm sizes continue to decrease.

While farm sizes are generally decreasing at the national level, there are areas within countries where farm sizes are increasing. In Thailand and the Philippines, these areas tend to be located near dynamic urban centres where the pull of the non-farm economy is so strong that the number of farms declines substantially. But even in these dynamic locations, the increase in farm size has to date been very small (Dawe, 2015). Thus, in summary, declining farm sizes are a key trend in farming systems in the region.

Small farm sizes can lead to three potentially serious problems. First, small farm sizes, other things being equal, mean lower household income for farmers. Second, small farm sizes might lead to inefficiencies in food production, endangering national food security. Third, small farms might have fewer incentives to adopt knowledge-intensive but environmentally friendly technologies. These are the farm, food and field problems identified by Christiaensen (2012).

A farmer with one hectare of land can, in nearly all cases, grow enough rice to meet the family's demand for staple food and achieve food security of the most basic kind. Indeed, with irrigated land that can grow two crops of rice a year, a typical farmer in most countries can make do with only a quarter of a hectare. But joining the rising middle class and being well-nourished is another matter entirely. In order to generate income that is adequate for purchasing a diverse diet, a mobile telephone, a motorcycle and sending children to school, a small rice field will not be enough – it will be essential to practise other more profitable crop/livestock/aquaculture systems, earn non-farm income, or both. There are constraints to farm production diversification, but such diversification (at the national level) is indeed taking place, as shown in the final section of Chapter 4.

Efficient farm production is important so that countries do not become too reliant on imports. While some degree of imports and exports is natural, ultimately it is a question of balance. The more competitive a nation's agriculture sector, the easier it is to achieve such balance. It is possible to control imports through trade policy as opposed to competitiveness, but there are clear disadvantages to using protectionist agricultural trade policies. Two key problems stand out. First, restricting imports via policy actions leads to higher domestic food prices, which makes it difficult for poor people without much land (in either urban or rural areas) to afford healthy, nutritious diets. Second, because food occupies as much as 70 percent of poor people's budgets, higher domestic food prices force unskilled wages to increase in compensation. In turn, labour-intensive industries become less internationally competitive, generate less employment and choke off a key pathway used by the poor to escape poverty. Thus, improving efficiency and competitiveness is essential for achieving national food security and escaping the middle-income trap.

Historically, small farms, especially in Asia, have been more efficient than large farms. Thus, at first glance, it is not obvious why ever-smaller farm sizes might create problems for the efficient production of food. The reason why smaller farms have been more efficient is that when wages were low (and mechanization was not cost-effective), smaller farms could be managed primarily through family labour, while larger farms would rely on hired labour. For many agricultural tasks, it is difficult to supervise hired labour, who have incentives to give less than full effort. Thus, because family labour has more incentive to work hard than hired labour, small farms were more efficient than large farms.

However, large farms also have some potential efficiency advantages over small farms. First, they have greater incentives to acquire knowledge and learn new farming techniques because for a fixed cost of learning the new technology or buying the new equipment, large farms can utilize this knowledge over a larger area. In the old era of stagnant technological development, this advantage was of no consequence. But today, in an era of rapid scientific progress, adoption of new technologies is crucial for efficient food production. Second, large farms may also have advantages in marketing in terms of producing large volumes of high quality and safe produce demanded by large retailers. Third, small farms may have difficulty adopting machines although the issue is more complex than often realized. Due to the existence of machinery service rental markets noted earlier, small farms do not need to buy machines – they can hire them instead. Machines can also be designed to operate on small parcels of land – for example, mechanization of rice harvesting is widespread in both China and Viet Nam, despite some of the smallest farm sizes in the region. Foster and Rosenzweig (2017) argue however that such redesigned machines, while serving a need in the context of rising wages, result in higher unit production costs.

There is some evidence that the efficiency advantage of small farms has been shrinking over time in several Asian countries (Otsuka, Liu and Yamauchi, 2016), although these estimates show that small farms are still relatively more efficient than large farms (just less so than in the past). Some studies show that larger farms are more efficient (e.g. Foster and Rosenzweig, 2017), while others show that smaller farms are equally productive and profitable, e.g. Rada, Wang and Qin (2015). Data from Japan and the Republic of Korea (Figure 6.9), two countries where wage rates are very high, show that the largest rice farms have the lowest production costs per tonne and the highest profits.

Figure 6.9 Costs of production per ton milled rice by farm size category, Republic of Korea, 2015

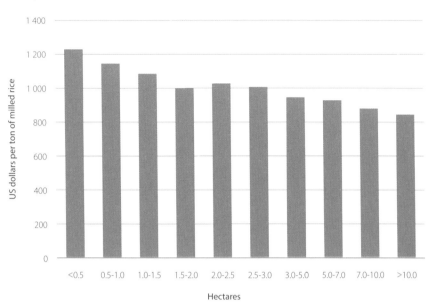

Source of raw data: KOSIS (2017)

If large farms are more efficient than small farms, a key question is just how large is the optimal size? Foster and Rosenzweig (2017) estimate that profits per unit of land in India reach a maximum at around 5 hectares – larger than many existing farms in India, but much smaller than typical farm sizes in the Americas and Australia. Data in Otsuka (2015) suggest that profits for rice farms in Japan are highest in farms between 5 and 10 hectares, although there are not many rice farms larger than those in Japan. Similarly, official government data from the Republic of Korea show that the largest farms have the lowest production costs per tonne, but the farms in that category are not that large, including farms that are just 10 hectares.

Small farms may also use non-labour inputs more intensively. The use of additional material inputs can increase yields up to a point, but it also poses dangers for the environment. In China, Wu *et al.* (2018) found that use of fertilizer and pesticides per hectare was greater in small farms than in large farms. Furthermore, despite slightly higher yields on small farms, nitrogen fertilizer runoff to the environment is greater from small farms due to lower nitrogen-use efficiencies.

Of course, the optimal farm size from a strictly economic point of view is not necessarily optimal from a societal point of view, as farms can serve as a safety net in times of crisis, and help prevent overly rapid migration to cities. Even if efficiency is the paramount objective, and if larger farms are more competitive and profitable (the evidence on this to date is still inconclusive in the overall regional context), there remains a major institutional question of how to consolidate farms into larger units while fully respecting the ownership and use rights of small family farmers. There remain many outstanding issues for further research in the realm of making small farms competitive while considering equity and societal objectives other than efficiency.

Summary

The incidence of rural poverty has declined in most countries of the region over the past 20 years. Part of this decline is due to the structural transformation described in Chapter 2 that has led farm households to rely increasingly on non-farm income as part of their livelihood strategy, especially in the face of declining farm sizes. Rural areas are also witnessing important demographic shifts. The population is ageing in many countries, especially the wealthier ones, but the trend is evident even in developing countries. In South Asia, male outmigration has led to women playing an increasingly important role in the agricultural labour force (often referred to as the feminization of agriculture).

Agricultural labour productivity (output per worker) has grown in most of the region (although the Pacific Islands are an exception), consistent with rising rural wages. Such increases in labour productivity are often realized through increased mechanization, and they can support improved livelihoods for farm households (along with increased non-farm household income). At the same time, mechanization results in loss of employment (especially for the rural landless) that is painful if the displaced workers cannot find alternative employment.

References

Allendorf, K. 2007. Do women's land rights promote empowerment and child health in Nepal? *World Development*, 35(11): 1975–1988. https://doi.org/10.1016/j.worlddev.2006.12.005

Bhandari, H. & Mishra, A.K. 2018. Impact of demographic transformation on future rice farming in Asia. *Outlook on Agriculture*, 47(2): 125–132. https://doi.org/10.1177/0030727018769676

Borras, S.M. 2005. Can redistributive reform be achieved via market-based voluntary land transfer schemes? Evidence and lessons from the Philippines. *Journal of Development Studies*, 41(1): 90–134. https://doi.org/10.1080/0022038042000276581

Bresciani, F., Feder, G., Gilligan, D.O., Jacoby, H.G., Onchan, T. & Quizon, J. 2002. Weathering the storm: The impact of the East Asian crisis on farm households in Indonesia and Thailand. *World Bank Research Observer*, 17(1): 1–20. https://doi.org/10.1093/wbro/17.1.1

Cho, A. 2017. The rapid rise of agricultural mechanization in Myanmar. *Presentation for the "Agriculture and Rural Transformation in Asia: Past Experiences and Future Opportunities" Conference, 13 December 2017, Bangkok.*

Christiaensen, L. 2012. The role of agriculture in a modernizing society: Food, farms and fields in China 2030. *Sustainable development - East Asia and Pacific Region discussion papers.* https://doi.org/doi.org/10.1596/26882

Davis, B., Di Giuseppe, S. & Zezza, A. 2017. Are African households (not) leaving agriculture? Patterns of households' income sources in rural Sub-Saharan Africa. *Food Policy*, 67: 153–174. https://doi.org/10.1016/j.foodpol.2016.09.018

Dawe, D. 2003. The future of large-scale rice farming in Asia. *Paper presented at the International rice conference on modern rice farming, Alor Setar, Malaysia, 13–16 October 2003.*

Dawe, D. 2005. Economic growth and small farms in Suphanburi, Thailand. *Presentation to the workshop on Agricultural Commercialization and the small farmer, 4–5 May 2005, Rome, FAO.*

Dawe, D. 2015. Agricultural transformation of middle-income Asian economies: Diversification, farm size and mechanization. ESA Working Paper No. 15-04. (also available at http://www.fao.org/publications/card/en/c/26f0548f-e90c-4650-9e17-9c378f6484dd/).

Deininger, K., Olinto, P. & Maertens, M. 2000. Redistribution, investment, and human capital accumulation: The case of Agrarian Reform in the Philippines. *Working Paper 28953.* (also available at http://documents.worldbank.org/curated/en/339161468775517919/Redistribution-investment-and-human-capital-accumulation-The-case-of-agrarian-reform-in-the-Philippines).

FAO. 2015. *Running out of time: The reduction of women's work burden in agricultural production.* F. Grassi, J. Landberg & S. Huyer, eds. (also available at http://www.fao.org/documents/card/en/c/da549560-cd7f-426c-9f6e-7228621cfbfd/).

FAO. 2016. *Addressing women's work burden: Key issues, promising solutions and way forward.* (also available at http://www.fao.org/family-farming/detail/en/c/472900/).

FAO. 2018. *FAOSTAT* [online]. www.fao.org/faostat/

FAO. 2018b. World Programme for the Census of Agriculture [online]. http://www.fao.org/world-census-agriculture/en/

Foster, A.D. & Rosenzweig, M.R. 2017. Are there too many farms in the world? Labor-market transaction costs, machine capacities and optimal farm size. *NBER Working Paper No. 23909.* https://doi.org/10.3386/w23909

Gulati, A., Saini, S. & Manchanda, S. 2017. Changing landscape of Indian farms: Holdings, labor, and machinery. *Presentation to World Food Policy Conference, 16–17 January 2017, Bangkok, Thailand.*

Hossain, M. 1996. Agricultural policies in Bangladesh: Evolution and impact on crop production. *State market and development: essays in honor of Rehman Sobhan.*

Kishore, A., Birthal, P.S., Joshi, P.K., Shah, T. & Saini, A. 2016. Patterns and drivers of dairy development in India: Insights from analysis of household and district-level data. *Agricultural Economics Research Review,* 29(1). https://doi.org/10.5958/0974-0279.2016.00014.8

Korean Statistical Information Service (KOSIS). 2017. *Rice production costs by size of plantedland* [online]. http://kosis.kr/statHtml/statHtml. do?orgId=101&tblId=DT_1EC0005&language=en&conn_path=I3

Lee, J. 2017. Aging of Korean farmers

Li, T.M. 2017. Intergenerational displacement in Indonesia's oil palm plantation zone. *The Journal of Peasant Studies,* 44(6): 1158–1176. https://doi.org/10.1080/03066150.2017. 1308353

Mandal, M.A.S. 2014. Agricultural mechanization in Bangladesh: Role of policies and emerging private sector. Paper presented at the NSD- IFPRI workshop on '*Mechanization and Agricultural Transformation in Asia and Africa: Sharing Development Experiences*', Beijing, China, 18–19 June 2014.

Moya, P.F., Bordey, F.H., Beltran, J.C., Manalili, R.G., Launio, C.C., Mataia, A.B., Litonjua, A.C. & Dawe, D.C. 2016. Costs of rice production. In F.H. Bordey, P.F. Moya, J.C. Beltran & D.C. Dawe, eds. *Competitiveness of Philippine rice in Asia,* pp. 99–117. International Rice Research Institute. (also available at http://www.philrice.gov.ph/wp-content/uploads/2016/08/Book_CPRA_22June2016_3.pdf).

Otsuka, K. 2015. *Future of small farms in emerging countries in Asia.*

Otsuka, K., Liu, Y. & Yamauchi, F. 2016. Growing advantage of large farms in Asia and its implications for global food security. *Global Food Security,* 11: 5–10. https://doi.org/10.1016/j.gfs.2016.03.001

Paris, T. 2017. Technological change and gender division of labor in irrigated rice villages: a comparative analysis across six Asian countries

Park, C.M.Y. & White, B. 2017. Gender and generation in Southeast Asian agro-commodity booms. *Journal of Peasant Studies,* 44(6): 1105–1112. https://doi.org/10.1080/0 3066150.2017.1393802

Poapongsakorn, N., Pantakua, K. & Wiwatvicha, S. 2016. The structural and rural transformation in selected Asian countries

Portilla, G.S. 2017. Land concessions and rural youth in Southern Laos. *The Journal of Peasant Studies,* 44(6): 1255–1274. https://doi.org/10.1080/03066150.2017.1396450

Qanti, S.R., Reardon, T. & Iswariyadi, A. 2017. Indonesian mango farmers participate in modernizing domestic. *Bulletin of Indonesian Economic Studies* (accepted December 2016).

Rada, N., Wang, C. & Qin, L. 2015. Subsidy or market reform? Rethinking China's farm consolidation strategy. *Food Policy,* 57: 93–103. https://doi.org/10.1016/j.foodpol.2015.10.002

Rahman, M., Saima, U. & Goni, A. 2015. Impact of maternal household decision-making autonomy on child nutritional status in Bangladesh. *Asia Pacific Journal of Public Health,* 27(5): 509–520. https://doi.org/10.1177/1010539514568710

Reardon, T., Chen, K., Minten, B. & Adriano, L. 2012. *The quiet revolution in staple food value chains: Enter the dragon, the elephant and the tiger.* ADB & IFPRI. (also available at http://www.ifpri.org/publication/quiet-revolution-staple-food-value-chains).

Reardon, T., Chen, K.Z., Minten, B., Adriano, L., Dao, T.A., Wang, J. & Gupta, S. Das. 2014. The quiet revolution in Asia's rice value chains. *Annals of the New York Academy of Sciences,* 1331: 1060–118. https://doi.org/10.1111/nyas.12391

Smith, T.C. 1988. Farm family by-employments in pre-industrial Japan. *The Journal of Economic History,* 29(4): 687–715. https://doi.org/10.1017/S0022050700071941

Verhofstadt, E., Maertens, M. & Swinnen, J. 2013. Smallholder participation in transforming agri-food supply chains in East Asia. Farmgate-to-market study on managing the agri-food transition in East Asia No. 3

White, B. 2018. *Rural youth today and tomorrow: Background paper (draft).* Rome, IFAD.

Wiggins, S. & Keats, S. 2014. Rural wages in Asia. London. (also available at https://www.odi.org/publications/8747-rural-wages-asia).

World Bank. 2018. *World development indicators* [online]. https://data.worldbank.org/products/wdi

Wu, Y., Xi, X., Tang, X., Luo, D., Gu, B., Kee, S., Vitousek, P.M. & Chen, D. 2018. Policy distortions, farm size, and the overuse of agricultural chemicals in China. PNAS, 125(27): 7010–7015. https://doi.org/10.1073/pnas.1806645115

CHAPTER 7

EVOLUTION OF FOOD SYSTEMS IN THE REGION

The Asia-Pacific region is dynamic and changing at a rapid pace. Due to the rapid changes outlined in Chapter 2, new considerations, described in Chapters 3 to 6, will increasingly shape food production, consumption and trade in the region. In moving forward, it will be essential to:

- Consider the environment and climate change (Chapter 3) in order to feed the world in the future at affordable prices, especially in light of a growing population. Agriculture affects the environment and climate change, which in turn affect our ability to produce food sustainably into the future. Market prices should incorporate the value of ecosystem services when feasible, but when this is difficult it will be necessary to have technologies and institutional innovations that provide these multiple services, both to feed the world and to create a healthy environment.

- Consider nutrition (Chapter 4) as a crucial part of human well-being and national competitiveness. Just a generation ago, people were substantially poorer and worried only about getting enough staple food (e.g. rice, wheat, roots and tubers) to eat. Now, growing numbers of people can afford diets that are more diverse and agriculture must meet these demands. Furthermore, for countries to compete effectively in the information age their citizens will need to achieve their full cognitive potential, which will require, among other things, nutritious diets at affordable prices. At the macro level, countries will also need to reduce the financial burdens placed on their health systems as they struggle to cope with an increased burden of obesity and non-communicable diseases.

- Consider the influence of value chains (Chapter 5) that are becoming increasingly complex, driven by more sophisticated consumer demands new to most farmers and policy-makers. The more sophisticated and complex demands, in particular the demand for increased convenience in food preparation, are especially noticeable in urban areas, which are growing rapidly. As a result, tackling malnutrition in urban areas will require, to some extent, solutions different to those implemented in rural areas.

- Consider farmer livelihoods (Chapter 6) within the context of overall rural development when designing interventions. Rising wages and the increasing importance of non-farm income in rural areas provide opportunities to achieve a greater degree of prosperity and manage the risks inherent in agricultural production. However, they also affect farmers' decision-making processes and the competitiveness of food production, although agricultural stakeholders often do not explicitly consider these factors as they are driven by changes outside the sector.

These four key considerations lie outside the domain of any single private sector entity, non-governmental organization (NGO) or government ministry. This fact underscores the increasingly important need for coordination across actors and sectors in planning and/or implementation, or at least an enhanced understanding of the important roles played by others ("think multisectorally, act sectorally" – Alderman *et al.*, 2013). This is consistent with a food systems approach that works to adopt integrated solutions to food system challenges based on holistic framing and deeper analysis of problems by a multidisciplinary team (FAO, 2018a).

It is important to note that there are often trade-offs between different objectives (e.g. an improved environment, better nutrition, more efficient value chains, greater farm incomes). For example, greater production of animal-source foods, which are rich in high-quality protein and bioavailable micronutrients, places stresses on the environment. Reducing greenhouse gas emissions through periodic drainage and flooding of paddy rice fields might reduce aquatic biodiversity. More irrigation will increase farmers' income but may accelerate groundwater depletion, leading to land subsidence or saltwater intrusion. Greater food imports can make food more affordable for consumers, thereby contributing to improved nutritional outcomes, but they also lower prices for farmers, thereby reducing farm household income (the classic food price dilemma – Timmer, Falcon and Pearson, 1983).

In addition to trade-offs between different objectives, different actors in society also often place different relative values on these objectives. There are also intertemporal trade-offs, with long-term benefits sometimes being realized only after incurring substantial short-term costs. Uncertainty around the exact nature of the impacts of any given intervention adds further to the difficulties. Despite these difficulties, it is important to recognize that these trade-offs exist and not assume that everyone at all times benefits from change. For example, recognition that trade-offs exist allows social safety nets to be helpful in pushing forward changes that are on balance helpful but harm certain groups, with political processes mediating these adjustments.

In addition to being dynamic, Asia and the Pacific is also a very diverse region with varying cultures and levels of per capita income across and within different countries. This heterogeneity across space and time implies that appropriate interventions in pursuit of a better environment, improved nutrition, more efficient value chains and greater farm household incomes will vary widely. However, it is useful to consider some general principles that are by no means exhaustive, but are likely to be appropriate at promoting pro-poor growth across a wide range of situations and have proven useful at promoting such growth in the past in a regional context (Timmer, 2004).

- Create an enabling environment at both macro- and micro-levels that promotes economic growth in an environmentally sustainable manner and simultaneously allows both men and women a range of opportunities to take initiative and control of their own lives;

- Create basic infrastructure (e.g. education, health care, roads, agricultural research, electrification, clean water and sanitation services, data and information and communication technology) that lowers the transactions costs for participating in markets;

- Promote development of the private sector and markets whenever possible, as the private sector (including smallholder farmers) is responsible for the bulk of farm (FAO, 2012a) and economy-wide investment, but regulate them when needed;

- Give special consideration to the interests of the poor and marginalized in both cities and the countryside, as they do not have a voice in markets (due to their low purchasing power) or in many other aspects of society;

- Give special consideration to youth, as they are the future; and

- Give special consideration to women, as they are playing an increasingly important role in food systems and could play an even larger role if they are given improved access to resources.

All of these principles are relevant across the entire food system from fork to farm, including consumption, marketing, production and the environment. While it is traditional to refer to the food system as 'farm to fork', market economies are demand-driven – therefore, the food system arguably starts more from the consumer than the farmer. For farmers' livelihoods to improve substantially, it is critical that smallholder farmers understand this reality and the demands that are being transmitted through increasingly sophisticated value chains. Governments must also create an enabling environment for smallholder farmers to participate in and benefit from these markets.

Food systems in the region are rapidly changing and there are many examples of how each of these principles are being applied to each of the four new considerations that are increasingly influencing the Asia-Pacific food economy. In order to give a flavour of these changes, this chapter will discuss a number of specific examples of such developments in the region, including their advantages and disadvantages and whether they might be more appropriate in some circumstances than in others. The choice of developments to discuss has not been based on a cost-benefit analysis; rather, the choices are purely meant to illustrate the increasing importance of the environment, nutrition, changing value chains and farmer livelihoods across a range of geographies. Some of the examples are new technologies, some are new institutions, some are new markets and some are old traditional interventions that are still important. Many of them will require coordination, at least in planning stages, across different sectors.

No single source can hope to provide specific recommendations on how to achieve a better environment, improved nutrition, more efficient value chains and greater farm household incomes. Thus, the main objective of the examples that follow is to foster a dialogue on possible ways to move forward in different regional contexts.

Environment and climate change

The solar-powered irrigation conundrum

Meeting the growing demand for food globally will require more water, land and energy (Hunter *et al.*, 2017). Of particular concern is the steady growth in the volumes of water needed to meet national food security goals. The agriculture sector already consumes up to 90 percent of overall water resources in some countries, and combined with growing demand from cities, industries and the environment, increasing water consumption in agriculture will continue to drive severe water scarcity across large parts of the region. A recent report from the OECD places Asia's irrigation-dependent food baskets in northwest India and north China as two of the world's top three hotspots in terms of water-related risks to food production (OECD, 2017).

Groundwater is of particular concern, given that Asia is the world's largest user of groundwater for irrigated agriculture (Shah *et al.*, 2007). Rural livelihoods and national food production have become increasingly reliant on groundwater, yet the levels of extraction now exceed natural replenishment rates in many areas. In northern India, for example, satellite-based estimates of groundwater storage-change indicate that groundwater is being depleted at a rate of approximately 17.7 ± 4.5 km^3/year, which is double the capacity of India's largest surface-water reservoir (Rodell, Velicogna and Famiglietti, 2009). While this rapid expansion of groundwater-irrigated crops across Asia has had a positive impact on rural incomes, poverty eradication and national food security (Mukherji *et al.*, 2009), all groundwater pumps that are powered by fossil fuels emit climate pollutants such as black carbon.

Solar powered irrigation systems (SPIS) can, therefore, play an important role in climate change mitigation, reducing carbon emissions in irrigated agriculture by replacing fossil fuels for power generation with a renewable energy source (FAO, 2018b). With this climate goal in mind, many governments are rolling out subsidy and incentive packages that are designed to increase the uptake of SPIS. In the Indian state of Rajasthan, for example, incentive schemes are providing as much as an 86 percent subsidy on the purchase of a single pump (Goyal, 2013).

Yet the positive impacts of solar powered irrigation systems in terms of reducing carbon emissions must be balanced against the very real risks these systems pose to fast-depleting groundwater supplies. Solar pumps offer approximately 2 300-2 500 hours a year of daytime energy (CGIAR, 2015). This provides farmers with a free and uninterrupted power source that can increase yield and incomes, but also risks exacerbating currently unsustainable levels of groundwater extraction. This risk is compounded by the fact that Asian groundwater economies are largely informal, meaning water extraction and use are neither monitored nor controlled by governments (Shah *et al.*, 2003).

Managing the urgent need to reduce carbon emissions from agriculture through climate smart technologies, without further incentivizing groundwater over-abstraction requires concerted policy attention and investment. First and foremost, water consumption by all sectors must be brought within sustainability limits. This necessitates a quantitative understanding of current water use (via water accounting) accompanied by evidence-based water allocation/quota systems. Establishing and enforcing such systems is vital and should be in place before governments subsidize new technologies that have the potential to accelerate groundwater extraction in the absence of government control over water resources. With a more accurate understanding of water resources obtained through consistent water accounting, governments will be in a position to better design their subsidy and other policies to ensure only suitable hydrogeological and social contexts are targeted.

Farmer adaptation to climate change

Changes in key climate variables over the longer time scales (decades or more) associated with climate change have been documented across the region. Effective adaptation responses include both incremental action now to address observed climate variability as well as investment in systemic and transformational measures to tackle more uncertain, potentially more catastrophic, future climate change scenarios (Vermeulen et al., 2013).

Climate-informed crop management is an incremental adaptation measure that has been widely adopted by farmers in a number of countries around the region. It includes practices such as adjusting planting dates to account for changes in seasonal weather patterns, use of different crop varieties with different growth duration or stress tolerance and use of supplementary irrigation (Abid et al., 2015; Abid, Schneider and Scheffran, 2016; Bastakoti et al., 2014; Bhatta and Aggarwal, 2015; Burnham and Ma, 2015; Chen, Wang and Huang, 2013; Dewi and Whitbread, 2017; Meng et al., 2014; Shaffril, Krauss and Samsuddin, 2018; Wood et al., 2014). For example, in Gujarat, India, farmers have adapted to delayed monsoon onset by switching to more drought-tolerant crops and/or delayed planting in combination with supplementary irrigation. In northeast China, gradual shifts to longer maturing maize varieties in tandem with shifts in climate suitability for the crop resulted in significant productivity improvements (Meng et al., 2014). It is important to note however, that additional requirements for water resulting from these changes may exacerbate water scarcity (Meng et al., 2016).

Wide adoption of climate-informed crop management is not a strictly autonomous phenomenon driven solely by farmers. Studies looking to understand the decision-making process of farmers in adopting on-farm innovations to address climate risks suggest that a number of enabling factors are required to facilitate the process. These factors include extension support services as well as availability of and access to more tolerant crop varieties and reliable weather and climate information (Bhatta *et al.*, 2017; Burnham and Ma, 2015; Shaffril, Krauss and Samsuddin, 2018; Wood *et al.*, 2014).

Agricultural extension and social support networks are a crucial factor behind documented examples of widespread adoption of climate-informed crop management practices (Abid *et al.*, 2015; Cui *et al.*, 2018; Wood *et al.*, 2014). Extension as well as social learning through peer exchange and farmer field schools were often found to be a mechanism for disseminating climate information and introducing alternative management options (Bastakoti *et al.*, 2014; Bhatta *et al.*, 2017; Hochman *et al.*, 2017). Government-supported extension programmes targeting a specific climate issue have in some cases been linked to farmer adoption of specific adaptation practices including the changing of planting dates or crop varieties (Bastakoti *et al.*, 2014).

The availability of improved, stress-tolerant crop varieties is essential for enabling climate-informed varietal selection. In many smallholder systems around the region, government and publicly-funded research institutes remain key providers of improved crop varieties such as submergence-tolerant rice (Atlin, Cairns and Das, 2017; Bhatta and Aggarwal, 2015). Future anticipated changes in climate will require continued innovation in the development of new crop varieties that are able to anticipate new magnitudes or types of climatic stress in the different agro-ecosystems across the region (Bhatta and Aggarwal, 2015; Mickelbart, Hasegawa and Bailey-Serres, 2015). Ensuring that plant breeders in national and international crop research institutes have the means to develop and distribute new, climate-adapted cultivars that will meet the needs of farmers, smallholders in particular, is crucial. For example, a collaboration between national and international research institutes along with private seed companies has led to the development of more than 50 heat stress-tolerant maize hybrids for licensing, release and deployment in the region (Cairns and Prasanna, 2018). Vigilance in the management of public seed production and distribution systems is also required to ensure that obsolete varieties are removed from circulation and that new varieties better adapted to prevailing conditions are made available to farmers on a regular basis (Atlin, Cairns and Das, 2017).

The provision of basic instruments for monitoring weather can lead to significant improvements in farmers' willingness and ability to manage climate risks and increase overall productivity (Bhatta *et al.*, 2017). Farmers who are provided access to climate forecasts use this information to make decisions about planting dates and the selection of crop varieties (Dewi and Whitbread, 2017; Wood *et al.*, 2014). In general, the level of investment and the quality and supply of weather and climate information services in developing countries in the region (and globally) is low (Georgeson, Maslin and Poessinouw, 2017; Rogers and

Tsirkunov, 2013), although there are exceptions. For example, a rapid assessment of national meteorological and hydrometeorological services in Southeast Asia found that Indonesia and the Philippines have relatively good quality observation networks in place as well as the human and institutional capacity to produce weather and climate information services for a range of users (Pieyns, 2014). Thus, systematic efforts to invest in climate observation and monitoring infrastructure to strengthen the provision of weather and climate information at the farmer level will be important enabling activities for government and other agriculture sector stakeholders to consider (Loboguerrero et al., 2017; Rogers and Tsirkunov, 2013)

Nutrition

Different approaches to reducing sugar content in food

Obesity and the incidence of non-communicable diseases (NCDs) such as diabetes is widespread in the Pacific – the ten countries with the highest rates of obesity are all Pacific Islands (NCD-RisC, 2018). NCDs are causing serious public health issues, as well as putting pressure on government budgets and diverting financial resources that could be used for investment. In Asia, the incidence of obesity is growing. Excessive intake of sugar is widely cited as one of the reasons behind these trends. Foods high in salt and fat and lack of exercise are also contributing.

Many countries in the region are now experimenting with sugar taxes, especially on bottled sugar-sweetened beverages, in an effort to reduce consumption. More than half of the 20 Pacific Island countries and territories monitored by the World Health Organization have taxes for sugar-sweetened beverages, with taxes typically ranging from 7 to 15 percent (Ives, 2017). The Philippines has also implemented a sugar tax, with higher rates on high fructose corn syrup (which is imported) to make the change more acceptable to domestic sugar producers (Jimenea, 2018). Other countries in the region are considering implementing such a tax, while some countries (e.g. Singapore, Hong Kong S.A.R.) have taken a different approach of working with private sector companies to reformulate their products so that they contain less sugar (Aravindan, 2017; Tang, 2018). Thailand has taken a hybrid approach, phasing in taxes over a period of time in order to encourage product reformulation that avoids the taxes (USDA, 2017).

Several studies (Osornprasop, 2017; Thow, Downs and Jan, 2014) found that taxes reduce consumption of targeted foods, but it is important to keep in mind the potential for consumer substitution into other foods that are not taxed (and perhaps not nutritious). For example, carbonated soft drinks are responsible for a substantial share of sugar consumption in the United States, but the importance of such beverages is lower in Asia, where, for example, tea sweetened at the point of sale is a major source. In this case, taxes on bottled sugar-sweetened beverages might encourage more consumption of sugar-sweetened tea, thus reducing the impact on total sugar consumption. Even in the United States, the beverage industry accounts for only about 30 percent of the total caloric sweetener market – most sugar and high fructose corn syrup is used in foods, not beverages (Cohen, DeFonseka and McGowan, 2017). Thus sugar taxes, if they are to help reduce sugar consumption, should consider a wide range of food and beverage products, although such a system may be administratively complex to implement.

Sugar taxes and product reformulation, if effective at reducing consumption, will have obvious negative implications for sugarcane farmers. In order to reduce these negative effects, governments can promote the production of alternative foods by designing a phase-out strategy and providing extension services and new knowledge to these farmers.

Given the different food preferences and cultural and political contexts across and within countries, there is no silver bullet solution to reducing sugar consumption. It will be important to experiment with multiple approaches and rigorously evaluate the outcomes.

Street food vendors and food safety

Consumer demand for increased convenience in food preparation, especially in urban areas, has led many people to eat more of their food away from home (see Chapter 5). Street food plays a major role in providing the food security and nutrition needs of lower income groups and also contributes to a city's attractiveness as a tourist destination. However, street food vending across large parts of the region is rife with unhygienic practices, largely due to a lack of education and awareness, leading to public health risks and unhealthy environmental effects due to waste and garbage. Because of their sheer numbers, informal street food vendors are often unregulated but nevertheless often subject to harassment from law enforcement and other agencies. Recognizing the importance of this unorganized sector, some governments in the region have taken steps to ensure stable premises or vending stations, maintain registers of street food vendors, provide a source of potable water and promote safe and hygienic street food as a sign of culinary heritage.

Singapore is one of the best-known examples of upgrading the environment that street food vendors operate in and increasing their knowledge of food safety issues. In addition to introducing a licensing and inspection system, Singapore built and established more than 100 hawker centres in the late 1970s and early 1980s and relocated street food vendors to these new facilities. Through zoning, rent control, logistical planning and subsidies for healthy meals, the government has ensured that safe, affordable and healthy food is available throughout the city. The government also implemented an 'educate and regulate' policy towards vendors and spent substantial sums of money to upgrade the hawker centres over time, for example by using infrastructure such as freezers that are shared by different vendors in the same centre (Jaffee *et al.*, 2018).

But it is not just Singapore that has been working with street food vendors. Bangkok, though much less regulated, requires that vendors undergo a two-day training course, which includes hygiene practices and garbage disposal. India has enacted federal legislation (Government of India, 2014) to register, improve skills, monitor quality and sustain the livelihoods of street vendors and has proposed a regional Codex standard for street food safety. Bangladesh has established a system of registration, training and monitoring in three cities and Dhaka is undertaking a full analysis and mapping of the circulation of food in the metropolitan area to facilitate food planning and identify the entry points for food safety risks.

Value chains and distribution systems

Solving coordination and aggregation problems in smallholder-based value chains

Agriculture in Asia and the Pacific is characterized by smallholder production. Nearly 90 percent of all smallholder farms of the world are in the region and farm size is declining even more over time (Chapter 6). Fragmented production and market interfaces give rise to high transaction costs and problems in matching supply with downstream consumer requirements, particularly those concerning food quality and safety. As a result, various stakeholders are working to solve the coordination and aggregation problems in smallholder-based value chains through contract farming, joint farming operations, shared services, farmer organizations or combinations of these models.

Contract farming is increasing in a wide range of countries in the region (FAO, 2013; Prowse, 2012), driven by the transformation of the food system described in previous chapters and further aided by government policies in some countries (e.g. China and Indonesia – Rehber, 2007). Contractual agreements seem to be more prominent in high-value agrifood chains (e.g. sugar, rubber, oil palm, tea, high-value fruits and vegetables, poultry, pork and aquaculture), while spot transactions continue to be the norm for undifferentiated staple crops. Contracts can also cross national borders. For example, many Laotian smallholder farmers have agreements with foreign private investors from China, Thailand and Viet Nam.

Another aggregation model that seems to be gaining ground in the region – sometimes in combination with contract farming – is joint- or block-farming. Such schemes consolidate small farms into a centrally-managed plot of land to take advantage of plantation-scale production or economies of scale. These are variations on the land rental markets described later, with the difference being that the owners of small plots continue to participate in the operation of the resultant larger farm, either working as labourers (lease agreements) or sharing input costs (growership agreements).

In the case of lease agreements, individual farmers lease out their small, adjacent landholdings to a central manager (e.g. agribusiness firm, cooperative) and become farm workers who earn income from both the lease rent and their salaries/wages. A case in point is the cooperative/joint farming initiative launched by the Nepal Agricultural Cooperative Central Federation (NACCFL), an apex organization that comprises over 800 agricultural cooperatives. Neighbouring farmers lease their land to the NACCFL for 15 years, and the management of the joint block (typically about 70 hectares) can help overcome the negative impacts of land fragmentation on input use, crop yield and production efficiency.

Under growership schemes, individual producers lease in land or consolidate their small landholdings into a viable land size to farm collectively. They share labour, input costs and financial returns, often according to a contract farming agreement with a buyer that contains stipulations on production, processing and marketing. Examples include Viet Nam for the production of hybrid rice seeds; the Philippines for banana in Davao del Norte and Compostela Valley, pineapple in South Cotabato, and sugarcane in Batangas (Pantoja, Alvarez and Sanchez, 2017); and the Indian states of Kerala and Telangana for a variety of high-value commercial crops (Agarwal, 2018).

Although there are challenges, these models show promise for helping to realize economies of scale within smallholder-based production systems and effectively address the aggregation and market-matching problems. Their suitability and inclusiveness vary depending upon the local context, which also affects the need for technical and legal support from various stakeholders.

Increasing sophistication of agrifood markets

Agrifood markets in the region are becoming more sophisticated along several dimensions: the development of differentiated products with new value-added attributes; new tools to verify the authenticity of such attributes; and innovative market arrangements to sell the new products.

The increasing sophistication of agrifood markets is partially due to consumers seeking higher-quality safe products in the wake of various food safety crises. They are switching to products with specific quality attributes, produced and certified according to certain principles such as organic agriculture, geographical indications (GIs) and ecosafe production systems.

Organic agriculture is growing fast in Asia. For example, the Indian state of Sikkim has achieved a total ban on the sale and use of chemical pesticides and is implementing a phase out of chemical fertilizers to become a fully organic state (FAO 2018d). Nevertheless, organic food remains a niche market in many countries (see Chapter 4), and per capita consumption is negligible compared to European or American consumers (Willer and Lernoud, 2018). However, these statistics offer just a partial picture – there are several institutional innovations that have the potential to create more demand by linking producers and consumers. Examples of such innovations include the formation of organic marketing clubs, community-supported agriculture (CSA) and Participatory Guarantee Systems (PGS) (Willer and Lernoud, 2018). In the CSA model, consumers subscribe to the harvest of a certain farm or group of farms, while the PGS model relies on the development of a locally-based system of quality assurance with a strong emphasis on social control and knowledge building. PGS initiatives can be found in Bangladesh, Cambodia, China, Fiji, India, Lao PDR, Nepal, Philippines, Sri Lanka, Thailand, Vanuatu and many other countries (FAO, 2018c).

GI products are another mechanism to connect producers and consumers more closely (FAO & SINER-GI, 2009). The GI concept incorporates specific characteristics (including quality and safety requirements), history and geography into the product in order "to value the unique" (Augustin-jean, Ilbert and Saavedra-Rivano, 2012). India's Darjeeling tea is just one example. Recently, the Association of Southeast Asia Nations (ASEAN) has been the most prolific region of the world in terms of GI registrations, with over 150 in the past decade, most of them dealing with food products (Rawat, 2017).

Innovative market arrangements to sell products with special attributes are also developing rapidly. These include direct sales to consumers using websites and/or smartphone apps, often catering for niche markets (allergen-free, pesticide-free, vegan/vegetarian, plastic-free and so forth). Fresh food e-commerce is developing rapidly in the region, with China leading the way and recording sales of USD 22.1 billion in 2017 (IResearch, 2018). This trend is most relevant for high-value food products: among foods purchased online, fruit is the most frequently bought; with dairy products and vegetables ranked second and third, respectively. Substantial investments in supply chains and logistics by e-commerce firms have enabled this growth.

There has also been rapid growth in databases and platforms that help consumers verify the authenticity of specific food attributes. Some cases in point are: the Indian Organic Integrity Database launched by the Food Safety and Standards Authority of India (Willer and Lernoud, 2018); IFOAM's map of international PGS initiatives (IFOAM, 2018); and ITC's Sustainability Map (ITC, 2017), an online platform that enables consumers, farmers, standard-setting organizations and the public sector to connect with others in the value chain. The registration of GIs can be found using the OriGIn (2018) compilation or regional databases such as the ASEAN GI database (ASEAN, 2018).

Farmers' livelihoods

Rental markets for land and services

As described in earlier chapters, the structural transformation brought on by rapid economic growth and migration to urban areas has led to an increased reliance of farm households on non-farm income. While these trends have many benefits for current and former farm households, it also means that many farms are either being abandoned or are now being managed by part-time farmers who have less interest in adopting new more productive and environmentally friendly technologies. The increasing age of farmers and declining farm sizes also give farmers less incentive to adopt new technologies.

Rental markets for land and services can help to alleviate these problems. China has established a large network of land transfer service centres that provide a range of services to farmers who want to rent out or rent in agricultural land (Huang, 2017). A key advantage of such land rental markets is that, for the farmer who rents in land, it will be easier to achieve economies of scale (see the discussion in Chapter 6). One potential disadvantage of land rental markets is that land might become more concentrated in the hands of large absentee landowners, making it politically difficult to implement in some circumstances.

An alternative to farm households renting out land is to rent in services. Thus, for example, farmers can rent the services of threshers, water pumps or combine harvesters without owning one themselves. In Bangladesh, while 72 percent of farmers use power tillers, only 2 percent of farmers own them (Ahmed, 2017). In Indonesia and the Philippines, the private sector provides mango spraying services in a similar manner (Qanti, Reardon and Iswariyadi, 2017). These service rental markets allow farmers to adopt the latest technologies and make their farms more productive, without having to learn the technology themselves or borrow money to purchase an expensive machine.

The key advantage of rental markets is that they allow individual farmers to choose whether they prefer to continue managing their own land, allow others to manage specific operations or allow someone else to manage it entirely on their behalf if they are too old or would prefer to spend more time in non-farm occupations. These land and service rental markets benefit the rentor (who is now earning more income in the non-farm sector, in addition to earnings from rental payments), the rentee (who is farming more land and earning more profits) and the service provider (who is growing their business). In particular, these markets can provide an important pathway for youth to gain access to land or to utilize their knowledge on information and communication technologies.

While such markets have many benefits, innovation to adapt machines to small farm sizes is still necessary and the public sector can support the development of prototype machinery suitable to local conditions. Governments can provide some limited tax incentives and enact regulations to smooth the functioning of credit markets, such as allowing machines to be used as collateral so that private banks are encouraged to lend to farmers or entrepreneurs. Financial subsidies for the purchase of machines work best if the farmer/entrepreneur has a substantial financial stake in promoting and delivering the machine rental service to other farmers. This is preferable to giving the machines away for free, as experience shows that such machines are often not repaired when they inevitably break down (Schmidley, 2014).

One key lesson from experience is that the government should not set ceiling prices for such rental services, or attempt to provide such services directly – the private sector will be more effective at this task. In Sri Lanka, the government established custom hiring centres in the early 1980s, but by the mid-1990s they had disappeared, having been outcompeted by private entrepreneurs (Samarasinghe, 2017).

Drones for agriculture

Farming communities have to adapt to climate change and other challenges. ICT-driven tools and technologies to enhance decision-making through accurate, reliable and timely information have an important role to play in this adaptation. One of the latest developments is the use of small, unmanned aerial vehicles (UAVs), commonly known as drones, for agriculture. They have a huge potential for improving spatial data collection and supporting evidence-based decision-making in agriculture. The use of drones in agriculture is extending at a brisk pace in crop production (precision farming), early warning systems and disaster risk reduction and management (DRRM), as well as in forestry, fisheries and wildlife conservation.

Precision crop farming combines sensor data and imaging with real-time data analytics to monitor and improve farm productivity through mapping spatial variability in the field. Data collected through drone sorties provide inputs to analytical models that can support soil health scans, monitoring of crop health, planning irrigation schedules and application of fertilizers. Drones fitted with infrared, multispectral and hyperspectral sensors can analyse crop health and soil conditions precisely and accurately, helping farmers to better manage their crops in real time. This will allow farmers to reduce input costs, increase yields, or both. Some of the main constraints to the greater use of drones are the cost of the drone itself and their limited flight range before the power supply is exhausted. The development of rental markets for drones (see the previous subsection) and improved battery storage technologies are likely to alleviate these constraints over time and make the technology affordable to smallholders.

In addition to precision farming, drones are increasingly used in the agricultural insurance sector, including in insurance claims forensics (Wadke, 2017). Drone imagery is very useful in giving an accurate estimate of losses, although there might be concerns over data privacy. Private companies are using drones to provide agriculture survey services to insurance companies and the state governments of Maharashtra, Gujarat, Rajasthan and Madhya Pradesh in India.

Drones are also being used to assist in DRRM efforts. The data collected by drones can provide rural communities with high-quality reliable advice to understand risks (e.g. related to landslides and erosion) and can assist the government in better planning disaster mitigation, relief and response. FAO has worked with the governments of Myanmar and the Philippines, including in remote areas (e.g. Rakhine and Chin states in Myanmar), to pilot the application of drone technology for these purposes.

Feed ration balancing for improved productivity and greater income

Livestock production is an important source of income for many farmers throughout the Asia-Pacific region and it helps to meet consumer demands for more diverse diets as well as providing a good source of nutrients for children. However, many farmers do not have access to scientific information on the composition of the local feeds that they use, nor are they always aware of the different nutrient requirements for different breeds of animals at different stages of the life cycle. These factors lead to a situation in which many animals receive inappropriate quantities of energy, protein and minerals. For example, a recent estimate for India's dairy sector found that two-thirds of milk cows received excess energy and protein, with a similar number being deficient in minerals (NDDB, 2017).

Imbalanced feeding can lead to lower meat or milk production, input costs that are higher than necessary and increased incidence of diseases, among others. Large private sector companies are well aware of such issues and devote considerable resources to balancing feed rations in an appropriate manner. Smallholders, however, have less knowledge and find it more difficult to optimize the content of feed rations. Government extension services have a role to play in overcoming this constraint. For example, the National Dairy Development Board (NDDB) of India has developed computer software that allows extension agents to advise farmers on how to create low-cost balanced rations using locally available feed ingredients, including both commercial feed as well as green fodder, crop residues and grasses and other feeds. The software takes account not only of animal characteristics such as body weight, milk yield, milk fat percentage and pregnancy status, but also the types of feed ingredients that are available in different agro-ecological regions.

This system has been successful at increasing milk production, lowering feed costs and increasing daily income by 10 to 15 percent, to the benefit of both male and female farmers (FAO, 2012b). As a result, it is estimated that 90 percent of India's dairy farms is using the output from this software, despite some initial misgivings by farmers regarding the ear-tagging of the cows (Searby, 2018). In addition to higher farm household income, ration balancing has also reduced enteric methane emissions by 12 to 15 percent per kilogram of milk.

One problem being experienced with this work is finding a continuing source of funding for the extension agents, as the initial push was funded through a special project. Finding money to support continuing farmer knowledge and education is likely to be a problem under many circumstances, but such support is important as scientific and technological advances mean that society's knowledge on agricultural production is continually improving and evolving.

References

Abid, M., Scheffran, J., Schneider, U.A. & Ashfaq, M. 2015. Farmers' perceptions of and adaptation strategies to climate change and their determinants: The case of Punjab province, Pakistan. *Earth System Dynamics*, 6: 225–243. Available at https://doi.org/10.5194/esd-6-225-2015

Abid, M., Schneider, U.A. & Scheffran, J. 2016. Adaptation to climate change and its impacts on food productivity and crop income: Perspectives of farmers in rural Pakistan. *Journal of Rural Studies*, 47(Part A): 254–266.

Agarwal, B. 2018. Can group farms outperform individual family farms? Empirical insights from India. *World Development*, 108: 57–73. Available at https://doi.org/10.1016/j.worlddev.2018.03.010

Ahmed, A.U. 2017. Patterns of farm mechanisation in Bangladesh. In M.A.S. Mandal, S.D. Biggs & S.E. Justice, eds. *Rural mechanisation: A driver in agricultural change and rural development*, pp. 119–134. Dhaka, Bangladesh, Institute for Inclusive Finance and Development (InM).

Alderman, H.H., Elder, L.K., Goyal, A., Herforth, A.W., Hoberg, Y.T., Marini, A., Ruel Bergeron, J., Saavedra Chanduvi, J., Shekar, M., Tiwari, S. & Zaman, H. 2013. *Improving nutrition through multisectoral approaches*. Washington DC, World Bank Group. (also available at http://documents.worldbank.org/curated/en/2013/01/17211210/improving-nutrition-through-multisectoral-approaches).

Aravindan, A. 2017. In health push, Singapore gets soda makers to cut sugar content. *Reuters Health News*, 21 August 2017. (also available at https://reut.rs/2fXLeeE).

Association of Southeast Asia Nations (ASEAN). 2018. *Geographical indications database* [online]. Available at http://www.asean-gidatabase.org/gidatabase/

Atlin, G.N., Cairns, J.E. & Das, B. 2017. Rapid breeding and varietal replacement are critical to adaptation of cropping systems in the developing world to climate change. *Global Food Security*, 12: 31–37. Available at https://doi.org/10.1016/j.gfs.2017.01.008

Augustin-jean, L., Ilbert, H. & Saavedra-Rivano, N., eds. 2012. *Geographical indications and international agricultural trade: The challenge for Asia*. Palgrave Macmillan. Also available at https://www.palgrave.com/gp/book/9780230355750#aboutAuthors

Bastakoti, R.C., Gupta, J., Babel, M.S. & van Dijk, M.P. 2014. Climate risks and adaptation strategies in the Lower Mekong River basin. *Regional Environmental Change*, 14: 207–219. Available at https://doi.org/10.1007/s10113-013-0485-8

Bhatta, G.D. & Aggarwal, P.K. 2015. Coping with weather adversity and adaptation to climatic variability: a cross-country study of smallholder farmers in South Asia. *Climate and Development*, 8(2): 1–13. Available at https://doi.org/10.1080/17565529.2015.1016883

Bhatta, G.D., Ojha, H.R., Aggarwal, P.K., Sulaiman, V.R., Sultana, P., Thapa, D., Mittal, N., *et al.* 2017. Agricultural innovation and adaptation to climate change: empirical evidence from diverse agro-ecologies in South Asia. *Environment, Development and Sustainability*, 19(2): 497–525. Available at https://doi.org/10.1007/s10668-015-9743-x

Burnham, M. & Ma, Z. 2015. Linking smallholder farmer climate change adaptation decisions to development. *Climate and Development*: 289–311. Available at https://doi.org/10.1080/17565529.2015.1067180

Cairns, J.E. & Prasanna, B.M. 2018. Developing and deploying climate-resilient maize varieties in the developing world. *Current Opinion in Plant Biology*, 45. Available at https://doi.org/10.1016/j.pbi.2018.05.004

Chen, H., Wang, J. & Huang, J. 2013. Policy support, social capital, and farmers' adaptation to drought in China. *Global Environmental Change*, 24(1): 193–202. Available at https://doi.org/10.1016/j.gloenvcha.2013.11.010

Cohen, E., DeFonseka, J. & McGowan, R. 2017. Caloric sweetened beverage taxes: A toothless solution? *The Economists' Voice*, 14(1). Available at https://doi.org/10.1515/ev-2017-0009

Consortium of International Agricultural Research Centers (CGIAR). 2015. Solar-powered water pumps offer green solution to irrigation. *Sunshine: India's new cash crop.* CGIAR Research Program on Water, Land and Ecosystems (WLE). Available at https://wle.cgiar.org/sunshine-india-new-cash-crop

Cui, Z., Zhang, H., Chen, X., Zhang, C., Ma, W., Huang, C., Zhang, W., et al. 2018. Pursuing sustainable productivity with millions of smallholder farmers. *Nature*, 555(7696): 363–366. Available at https://doi.org/10.1038/nature25785

Dewi, E.R. & Whitbread, A.M. 2017. Use of climate forecast information to manage lowland rice-based cropping systems in Jakenan, Central Java, Indonesia. *Asian Journal of Agricultural Research*, 11(3): 66–77. Available at https://doi.org/10.3923/ajar.2017.66.77

FAO. 2012a. The state of food and agriculture 2012: Investing in agriculture for a better future. (also available at http://www.fao.org/publications/sofa/2012/en/).

FAO. 2012b. Balanced feeding for improving livestock productivity: Increase in milk production and nutrient use efficiency and decrease in methane emission. *FAO Animal Production and Health Paper No.173.* (also available at http://www.fao.org/docrep/016/i3014e/i3014e00.htm).

FAO. 2013. *Contract farming for inclusive market access.* C.A. da Silva & M. Rankin, eds. Rome. (also available at http://www.fao.org/3/a-i3526e.pdf).

FAO. 2018a. Food systems and value chains: definitions and characteristics. In: *Production and resources: developing sustainable food systems and value chains for climate-smart Agriculture* [online]. http://www.fao.org/climate-smart-agriculture-sourcebook/production-resources/module-b10-value-chains/chapter-b10-2/en/

FAO. 2018b. *The benefits and risks of solar-powered irrigation – a global overview.* H. Hartung & L. Pluschke, eds. FAO & GiZ. (also available at http://www.fao.org/3/i9047en/I9047EN.pdf).

FAO. 2018c. Participatory Guarantee Systems (PGS) for sustainable local food systems. (also available at http://www.fao.org/3/I8288EN/i8288en.pdf).

FAO. 2018d. Sikkim, India's first "fully organic" state wins FAO's Future Policy Gold Award. (also available at http://www.fao.org/india/news/detail-events/en/c/1157760/).

FAO & SINER-GI. 2009. *Linking people, places and products: A guide for promoting quality linked to geographical origin and sustainable geographical indications.* Second edition. E. Vandecandelaere, F. Arfini, G. Belletti & A. Marescotti, eds. Available at http://www.fao.org/sustainable-food-value-chains/library/details/en/c/266257/

Georgeson, L., Maslin, M. & Poessinouw, M. 2017. Global disparity in the supply of commercial weather and climate information services. *Science Advances*, 3(e1602632). Available at https://doi.org/10.1126/sciadv.1602632

Government of India. 2014. The Street Vendors (Protection of Livelihood and Regulation of Street Vending) Act, 2014. Available at https://www.india.gov.in/street-vendors-protection-livelihood-and-regulation-street-vending-act-2014

Goyal, D.K. 2013. Rajasthan solar water pump programme: Sustainable future for farmers. *Akshay Urja Renewable Energy*, 7(2&3): 10–18. Available at https://mnre.gov.in/file-manager/akshay-urja/september-december-2013/EN/10-18.pdf

Hochman, Z., Horan, H., Reddy, D.R., Sreenivas, G., Tallapragada, C., Adusumilli, R., et al. 2017. Smallholder farmers managing climate risk in India: 1. Adapting to a variable climate. *Agricultural Systems*, 150: 54–66. Available at https://doi.org/10.1016/j.agsy.2016.11.007

Huang, J. 2017. *Land transaction service centers in China: An institutional innovation to facilitate land consolidation.* Beijing.

Hunter, M.C., Smith, R.G., Schipanski, M.E., Atwood, L.W. & Mortensen, D.A. 2017. Agriculture in 2050: Recalibrating targets for sustainable intensification. *BioScience,* 67(4): 386–391. Available at https://doi.org/10.1093/biosci/bix010

International Federation of Organic Agriculture Movements (IFOAM). 2018. *Map of participatory guarantee systems worldwide* [online]. Available at https://pgs.ifoam.bio/

International Trade Center (ITC). 2017. Routes to inclusive and sustainable trade. *International Trade Forum(4).* Available at http://www.intracen.org/uploadedFiles/Common/Content/TradeForum/Trade_Forum_4_2017.pdf

IResearch. 2018. *China's fresh food e-commerce consumption report.* Available at http://www.iresearchchina.com/content/details8_41001.html

Ives, M. 2017. As obesity rises, remote Pacific Islands plan to abandon junk food. *New York Times,* 19 February 2017. Available at https://www.nytimes.com/2017/02/19/world/asia/junk-food-ban-vanuatu.html

Jaffee, S., Henson, S., Unnevehr, L., Grace, D. & Cassou, E. 2018. *The safe food imperative: Accelerating progress in Low- and Middle-Income countries.* Washington, DC, World Bank Group. Available at https://openknowledge.worldbank.org/handle/10986/30568

Jimenea, L. 2018. Soda makers eye more local sugar input. The Philippine Star, 11 January 2018. Available at http://po.st/Kp1kTp

Loboguerrero, A.M., Birch, J., Thornton, P., Meza, L., Sunga, I., Bong, B.B., Rabbinge, R., *et al.* 2017. *Feeding the world in a changing climate: An adaptation roadmap for agriculture.* Rotterdam and Washington, DC, The Global Commission on Adaptation. Available at https://hdl.handle.net/10568/97662

Meng, Q., Chen, X., Lobell, D.B., Cui, Z., Zhang, Y., Yang, H. & Zhang, F. 2016. Growing sensitivity of maize to water scarcity under climate change. *Scientific Reports,* 6(19605). Available at https://doi.org/10.1038/srep19605

Meng, Q., Hou, P., Lobell, D.B., Wang, H., Cui, Z., Zhang, F. & Chen, X. 2014. The benefits of recent warming for maize production in high latitude China. *Climatic Change,* 122(1–2): 341–349. Available at https://doi.org/10.1007/s10584-013-1009-8

Mickelbart, M. V., Hasegawa, P.M. & Bailey-Serres, J. 2015. Genetic mechanisms of abiotic stress tolerance that translate to crop yield stability. *Nature Reviews Genetics,* 16(4): 237–251. Available at https://doi.org/10.1038/nrg3901

Mukherji, A., Facon, T., Burke, J., Fraiture, C. de, Faurès, J.M., Füleki, B., Giordano, M., *et al.* 2009. *Revitalizing Asia's irrigation to sustainably meet tomorrow's food needs.* International Water Management Institute (IWMI), FAO. Available at http://www.iwmi.cgiar.org/Publications/Other/PDF/Revitalizing Asia%27s Irrigation.pdf).

National Dairy Development Board (NDDB). 2017. Implementing a Ration Balancing Programme (RBP) in India. (also available at https://dairysustainabilityframework.org/wp-content/uploads/2017/11/Winning-Poster.pdf).

NCD Risk Factor Collaboration (NCD-RisC). 2018. *NCD-RisC data and publications* [online]. Available at http://ncdrisc.org/

Organisation for Economic Co-operation and Development (OECD). 2017. *Water risk hotspots for agriculture.* OECD Publishing. Available at http://www.oecd-ilibrary.org/agriculture-and- food/water-risk-hotspots-for-agriculture_9789264279551-en

Organization for International Geographical Indications (oriGIn). 2018. *Online database of geographical indications.* Available at https://www.origin-gi.com

Osornprasop, S. 2017. *Tonga NCD-related taxation policy assessment: Preliminary findings from government data and baseline household/retail surveys.* Paper presented at the Asia and the Pacific Symposium on Sustainable Food Systems for Healthy Diets and Improved Nutrition, 10–11 November 2017, Bangkok, Thailand.

Pantoja, B.R., Alvarez, J.V. & Sanchez, F.A. 2017. *Assessment of agribusiness venture arrangements and sugarcane block farming for the modernization of agriculture.* Philippine Institute for Development Studies. Available at https://pidswebs.pids.gov.ph/CDN/PUBLICATIONS/pidsdps1735.pdf

Pieyns, S. 2014. *Assessment on the state of hydrological services and proposals for improvement. Phase I: Rapid global assessment proposals for further in-depth capacity assessment in selected countries.* Washington, DC, World Bank.

Prowse, M. 2012. Contract farming in developing countries: a review. *A savoir,* 12. Paris, AFD. Available at https://www.afd.fr/en/contract-farming-developing-countries-review

Qanti, S.R., Reardon, T. & Iswariyadi, A. 2017. Indonesian mango farmers participate in modernizing domestic. *Bulletin of Indonesian Economic Studies* (accepted December 2016).

Rawat, P. 2017. Unfolding geographical indications of India: A brief introduction. *International Journal of Advanced Research and Development,* 2(5): 497–508. Available at http://www.advancedjournal.com/archives/2017/vol2/issue5

Rehber, E. 2007. *Contract farming: Theory and practice.* The ICFAI University Press. Available at http://ageconsearch.umn.edu/record/259070/files/Rehber1.pdf

Rodell, M., Velicogna, I. & Famiglietti, J.S. 2009. Satellite-based estimates of groundwater depletion in India. *Nature,* 460(20): 999–1003. Available at https://doi.org/10.1038/nature08238

Rogers, D.P. & Tsirkunov, V.V. 2013. *Weather and climate resilience: Effective preparedness through national meteorological and hydrological services.* Washington, DC, World Bank. Available at http://documents.worldbank.org/curated/en/308581468322487484/Weather-and-climate-resilience-effective-preparedness-through-national-meteorological-and-hydrological-services

Samarasinghe, M. 2017. *Custom hiring: Impact on paddy cultivation in Sri Lanka.* Agfour Engineering Services.

Schmidley, A. 2014. Measures for reducing post-production losses in Rice. In D. Dawe, S. Jaffee & N. Santos, eds. *Rice in the shadow of skyscrapers: Policy choices in a dynamic East and Southeast Asian setting,* pp. 67–72. Rome, FAO.

Searby, L. 2018. India: Ration balancing boosts dairy producer profitability. *FeedNavigator.* Available at https://www.feednavigator.com/Article/2018/08/14/India-Ration-balancing-boosts-dairy-producer-profitability?

Shaffril, H.A.M., Krauss, S.E. & Samsuddin, S.F. 2018. A systematic review on Asian's farmers' adaptation practices towards climate change. *Science of the Total Environment,* 644: 683–695. Available at https://doi.org/10.1016/j.scitotenv.2018.06.349

Shah, T., Burke, J., Villholth, K., Angelica, M., Custodio, E., Daibes, F., Hoogesteger, J., *et al.* 2007. Groundwater: a global assessment of scale and significance. In D. Molden, ed. *Water for food, water for life: A comprehensive assessment of water management in agriculture,* pp. 395–423. Earthscan & IWMI. Available at http://dspace.ilri.org/handle/10568/35042

Shah, T., Roy, A.D., Qureshi, A.S. & Wang, J. 2003. Sustaining Asia's groundwater boom: An overview of issues and evidence. *Natural Resources Forum,* 27: 130–141. Available at https://onlinelibrary.wiley.com/doi/pdf/10.1111/1477-8947.00048

Tang, A.S.P. 2018. Reduction of sugars in prepackaged foods and beverages. *Food Safety Focus* 139 . Available at https://www.cfs.gov.hk/english/multimedia/multimedia_pub/multimedia_pub_fsf_139_02.html).

Thow, A.M., Downs, S. & Jan, S. 2014. A systematic review of the effectiveness of food taxes and subsidies to improve diets: Understanding the recent evidence. *Nutrition Reviews*, 72(9): 551–565. Available at https://doi.org/10.1111/nure.12123

Timmer, C.P. 2004. The road to pro-poor growth: The Indonesian experience in regional perspective. *Bulletin of Indonesian Economic Studies*, pp. 177–207.

Timmer, P., Falcon, W.P. & Pearson, S.R. 1983. *Food policy analysis*. Baltimore & London, The John Hopkins University Press. Available at http://documents.worldbank.org/curated/en/308741468762347702/Food-policy-analysis

United States Department of Agriculture (USDA). 2017. Thai Excise Department implements new sugar tax on beverages. *GAIN Report* (TH7138). Available at https://gain.fas.usda.gov/Recent GAIN Publications/Thai Excise Department Implements New Sugar Tax on Beverages_Bangkok_Thailand_10-17-2017.pdf

Vermeulen, S.J., Challinor, A.J., Thornton, P.K., Campbell, B.M., Eriyagama, N., Vervoort, J.M., Kinyangi, *et al.* 2013. Addressing uncertainty in adaptation planning for agriculture. *PNAS*, 110(21): 8357–8362. Available at https://doi.org/10.1073/pnas.1219441110

Wadke, R. 2017. *Insurers deploy drones to check claims by farmers* [online]. Available at https://www.thehindubusinessline.com/economy/agri-business/insurers-deploy-drones-to-check-claims-by-farmers/article9583909.ece

Willer, H. & Lernoud, J., eds. 2018. *The world of organic agriculture: Statistics and emerging trends 2018*. FiBL & IFOAM. Available at https://www.ifoam.bio/en/news/2018/06/19/download-world-organic-agriculture-2018-free

Wood, S.A., Jina, A.S., Jain, M., Kristjanson, P. & DeFries, R.S. 2014. Smallholder farmer cropping decisions related to climate variability across multiple regions. *Global Environmental Change*, 25: 163–172. https://doi.org/10.1016/j.gloenvcha.2013.12.011